M000309589

NavCivGuide

Titles in the Series

American Admiralship
The Bluejacket's Manual
Career Compass
The Chief Petty Officer's Guide
Command at Sea
Dictionary of Modern Strategy and Tactics
Dictionary of Naval Abbreviations
Dictionary of Naval Terms
Division Officer's Guide
Dutton's Nautical Navigation
Ethics for the Junior Officer
Farwell's Rules of the Nautical Road
Naval Ceremonies, Customs, and Traditions
The Naval Institute Almanac of the U.S. Navy
The Naval Institute Guide to Naval Writing
The Naval Officer's Guide
Naval Shiphandler's Guide
Principles of Naval Weapon Systems
The Professional Naval Officer: A Course to Steer By
Reef Points
A Sailor's History of the U.S. Navy
Watch Officer's Guide

Forthcoming

Guide to Naval Architecture
Guide to Naval Writing
The Mariner's Guide to Meteorology
The Mariner's Guide to Oceanography
The Newly Commissioned Officer's Guide
The Petty Officer's Guide

The U.S. Naval Institute Blue & Gold
Professional Library

For more than a hundred years, U.S. Navy professionals have counted on specialized books published by the Naval Institute Press to prepare them for their responsibilities as they advance in their careers and to serve as ready references and refreshers when needed. From the days of coal-fired battleships to the era of unmanned aerial vehicles and laser weaponry, such perennials as *The Bluejacket's Manual* and the *Watch Officer's Guide* have guided generations of Sailors through the complex challenges of naval service. As these books are updated and new ones are added to the list, they will carry the distinctive mark of the Blue & Gold Professional Library series to remind and reassure their users that they have been prepared by naval professionals and meet the exacting standards that Sailors have long expected from the U.S. Naval Institute.

BLUE & GOLD
PROFESSIONAL LIBRARY

Naval Institute Press
291 Wood Road
Annapolis, MD 21402

© 2008 by Thomas J. Cutler
All rights reserved. No part of this book may be reproduced or utilized in any form or by any means, electronic or mechanical, including photocopying and recording, or by any information storage and retrieval system, without permission in writing from the publisher.

Library of Congress Cataloging-in-Publication Data
Cutler, Thomas J., 1947–
   Navcivguide : a handbook for civilians in the United States Navy / Thomas J. Cutler.
      p. cm. — (U.S. Naval Institute Blue & Gold professional library)
   Includes index.
   ISBN-13: 978-1-59114-155-6 (alk. paper)
   ISBN-10: 1-59114-155-9 (alk. paper)
   1. United States. Navy—Civilian employees—Handbooks, manuals, etc.   I. Title.
VB183.C87 2008
359.00973—dc22   2006027116

Printed in the United States of America on acid-free paper ♾

14 13 12 11 10 09 08   9 8 7 6 5 4 3 2
First printing

All ribbons, pins, and rating badges courtesy of *All Hands* magazine.

NavCivGuide

A Handbook for Civilians
in the United States Navy

THOMAS J. CUTLER

Naval Institute Press
Annapolis, Maryland

*To all who serve—
in and out of uniform—
in war and peace.*

Contents

Illustrations

FIGURES

PHOTOGRAPHS

TABLES

Acknowledgments

I am honored to be able to include this impressive list of people who selflessly assisted me in writing this book. They vetted the chapters as they were written, answered my many queries, provided creative advice, and gave me great encouragement as I wrote. Were I to list each one's contributions in any detail, this would be a much longer book. Their only reward is my deep personal gratitude and the knowledge that by helping those who serve, they are contributing to the strength of this great Navy of ours.

LCDR Youssef Aboul-Enein, MSC, USN—Middle East Policy Advisor, Office of the Secretary of Defense for International Security Affairs

Patricia Adams—Deputy Assistant Secretary of the Navy (Civilian Human Resources)

Michael Beyrle Sr.—Asst. Operations Officer, Naval Support Activity Washington, Washington Navy Yard

Paul G. Boundy—Human Resource Specialist, Navy Personnel Command

Deborah W. Cutler—Author, editor, creative consultant, critic, and loving wife

Dr. Bruce A. Elleman—Associate Professor, Maritime History Department, U.S. Naval War College

Peter A. Fitch—Workforce Development Specialist, Office of Civilian Human Resources, Department of the Navy

Dr. Alan B. Flanders—Historian, Center for Naval Leadership

COL Charles Gentile, AUS (Ret.)—Former counterintelligence officer and Staff Assistant and Tutor at the Schuler School of Fine Arts in Baltimore, Maryland

Mary Glotfelty—Director, Civilian Workforce Development Division, Office of Civilian Human Resources

Dr. Frances G. Holt—Executive Director, Navy Munitions Command

Dr. Janice H. Laurence—Director, Research and Analysis Office of the Undersecretary of Defense, Personnel, and Readiness

Dr. Edward J. Marolda—Senior Historian and Chief, Histories and Archives Division, Naval Historical Center

John H. O'Brien—Director of Periodicals, U.S. Naval Institute

Jeb C. Parr—New Employee Program Manager, Puget Sound Naval Shipyard and Intermediate Maintenance Facility

Michele Roberts—Staff Director/HR Specialist to the Deputy Assistant Secretary of the Navy (Civilian Human Resources)

LCDR William Rogers—Future Force Staff, Assistant Commandant for Human Resources, U.S. Coast Guard

Jill M. Thomas—Administrative Officer, Naval Historical Center

Jon T. Youngdahl—Special Events Coordinator, Naval Support Activity, Washington

Chris Onrubia, Jessica Schultheis, and Mary Svikhart deserve a special thanks as the talented, dedicated, and "unsung" professionals who spent many hours of editing, proofing, and designing this book.

Introduction

USS *Yorktown* limped up the Pearl Harbor channel, a telltale oil slick painting her wake as she went. Two near misses and a direct hit from Japanese aircraft had done considerable damage to this aircraft carrier and it seemed certain that she would have to sit out the war for a number of weeks—perhaps even months—while her internal organs were put back in order.

But other forces were at work. Huge forces, in fact. A large Japanese armada—the largest yet assembled in the Pacific War—was headed for an island northwest of Hawaii—an atoll named Midway, diminutive in area but gigantic in strategic significance. Should they capture this American outpost, the Japanese might well drive the U.S. Navy into a retreat all the way back to the West Coast of America.

Only three aircraft carriers and a handful of cruisers and destroyers were left to meet this gargantuan enemy force, and *Yorktown* was one of those three carriers. If the U.S. Fleet were to have any chance at all in the coming showdown, this limping carrier would have to be included.

As she entered a drydock, Admiral Chester Nimitz, Commander of the Pacific Fleet, wearing hip boots like the others, sloshed around in seawater deep in the ship's innards, dodging hanging cables and peering at buckled plates and jagged metal as the repair experts tallied up the weeks of work needed. Nimitz turned to the supervisor and quietly but firmly said, "We *must* have the ship back in three days."

Civilian yard workers swarmed aboard armed with a different arsenal of war—hammers, acetylene torches, and the like—and soon the ship echoed with a cacophony of frantic but purposeful activity.

Bill Bennett, a large, imposing man—some defined him as a "formidable force"—was the supervisor of the 150 shipfitters of Shop 11-26. From the moment his workers boarded *Yorktown*, Bennett was clearly in charge. Under his direction, burners and riggers and welders set about their work, cutting out the wreckage that was in the way and removing it, creating replacements for needed structural members, and quickly welding them into place. They didn't refer to plans or even create sketches—there was simply no time for that. It was a time for improvisation and creativity, not standard procedure. Although they did not know any of the details, these workers knew that the Japanese were on their way and that every second counted.

Occasionally sucking on an orange to keep himself going, Bennett moved among his men, assessing their work and their condition. When one looked to be on the verge of dropping, Bennett would send him topside for a sandwich and a breath of air. Before long, the shipfitter would go back to face more hours of intense heat, suffocating fumes, and backbreaking labor.

It was the same everywhere. Like Bill Bennett, Ellis Clanton, Chief Quarterman of Shop 31, worked for three days straight, restoring *Yorktown*'s elevators and arresting gear to working order. Fred Rodin and his mates wrapped miles of insulation around mazes of pipes in the carrier's huge firerooms and enginerooms. Working around the clock in temperatures sometimes reaching 120°, shipyard workers swarmed about the ship, hammering and cutting and welding and wiring in an eerie world of pulsating light, choking smoke, pungent fumes, and a racing clock. Men collapsed from exhaustion, were hauled topside until they revived, and then returned to the hell below. One man was found asleep on a scaffold, his cutting torch still in hand.

Three days later, the resurrection was complete. Though *Yorktown* was nowhere near fully restored to her prebattle condition, the workers had successfully—incredibly—transformed her from useless hulk to a fighting ship. She steamed down the channel, headed for sea and a "rendezvous with destiny," civilian workers spilling from her insides into small boats alongside as she went.

What followed was arguably the greatest naval battle in history, certainly one that turned the tide of the Pacific War. The courage and sacrifice of the men who fought the battle of Midway is legendary and little short of miraculous, but the miracle began when others—civilians on the Navy payroll—fought exhaustion and the clock to do the seemingly impossible. The victory at Midway is theirs as well.

During that same war, some five thousand miles away, one woman described her wartime experience as "a solid year on the graveyard shift so as to be home with the kids during the day . . . indigestible lunches, long hours, and no promise of a future after the war—all for miserably low wages." Yet she and twenty-six thousand other men and women worked in the Washington Navy Yard gun factory, not only manufacturing gun barrels and other weapon components, but also serving as part of the "nerve center" of the Navy's ordnance design and testing program, furnishing blueprints and specifications to civilian gun factories around the nation and thereby ensuring ordnance standardization and ultimately contributing to the hard-won victory at sea in World War II.

And standing before the flagstaff in that same Washington Navy Yard on 1 June 1861, while the nation was being torn apart by civil war, a slave by the name of Michael Shiner took an oath of allegiance to the United States. Even though he would not be freed for another ten months, he and others like him contributed to the eventual Union triumph by working on the defenses of the strategically vital Navy yard and providing much-needed provisions to the U.S. Capitol in the harrowing early days of the conflict, when a Confederate assault seemed imminent. Shiner had witnessed the British burning of Washington nearly fifty years earlier in the War of 1812 and was determined not to see that happen again. As both a slave and a free man, Shiner spent a total of fifty-two years working for the Navy.

Howard Lorenzen did not spend a half-century serving the Navy as Michael Shiner had done, but in his thirty-three years with the Naval Research Laboratory, he made many significant contributions to the nation's defense. Known to many as the "father of electronic warfare," Lorenzen began his career with the laboratory in July 1940, working with radar and what he then called "radio-countermeasures." His work continued through three separate wars, and he was a key player in many of the electronic breakthroughs that helped the United States survive and eventually prevail in the Cold War.

In 1960, he headed a team that was instrumental in deploying the first electronic intelligence satellite system, getting the first one into space just four days after the Soviet Union had successfully countered American manned reconnaissance by shooting down a U-2 spy plane. That system collected important data that caused a major shift in U.S. strategy and led to the development of the American intercontinental ballistic missile (ICBM) system both on land and at sea.

Connie Grigor has been the secretary for the U.S. Naval Academy's history department for more than two decades. She is the behind-the-scenes force that keeps things moving and the glue that holds things together in one of the many academic departments at the school. She sees to the myriad needs of more than fifty professors, civilian and military, who prepare the young midshipmen to be officers in the Navy and Marine Corps. As chairmen of the department serve their terms and move on, as officers teach for a few years and then go back to the Fleet, as civilian professors retire to focus at last on their research and writing, it is Mrs. Grigor who keeps the watch, ordering textbooks, maintaining schedules, preparing curricula, and tending to the never-ending tides of paper that ebb and flow through the great doors of Sampson Hall. The Naval Academy's history department depends upon many people to function, but they *all* depend upon her.

Stories like these could fill this entire book, but the point should already be evident. The success of the U.S. Navy in its more than two centuries of existence is due not only to the essential contributions of Sailors on active duty and in the Reserve, but also to that other vital arm of the Total Force triad, the civilians who have worked as part of the Navy since its earliest days. In March 2004, Secretary of the Navy Gordon England told the U.S. Senate, "A large part of the credit for the Navy's outstanding performance goes to our civilian workforce. These experienced and dedicated craftspeople, researchers, supply and maintenance specialists, computer experts, service providers and their managers are an essential part of our Total Naval Force concept."

But these essential people come into the Navy with little or no understanding of a culture that is in many ways foreign to them. Before they go to work in the Fleet, active and reserve Sailors go to Boot Camp or officer candidate school to prepare them for their new (unique) occupation. And the Navy has long provided *The Bluejacket's Manual* to incoming Sailors to serve as an introduction and as a continuing reference so that they will feel more comfortable in a new and otherwise alien world, where floors suddenly become decks, where 1337 is a time in the here and now instead of a date from ancient history, and where uniforms are anything but! Yet the civilians who enter this alien world are not prepared in a similar manner.

Though it is impractical to send all civilian workers to a centralized indoctrination course, it *is* possible to provide a common reference, specially designed to acquaint civilians with this very special world they have entered. This book is that common reference, designed specifically for you who make up that vital third leg of the Total Force triad, who, like the Sailors in the Fleet, serve the nation and the Navy, who need and deserve help in understanding where you are and what it is all about. And though this book certainly cannot tell you everything you need to know to do your job in the Navy, it is a good start, a good way to take much of the mystery out of it all.

All organizations and occupations have their own idiosyncrasies, and a big step toward "fitting in" has always been learning how to "talk the talk and walk the walk." Like *The Bluejacket's Manual*, this guide provides the words and steps needed to serve as an introduction for new employees and as a ready reference for veteran workers. In the pages that follow, questions such as these will be answered:

- How do I answer a letter from someone who signs her name AB3 Sara Jones?
     See "Military Titles"
- Who is my boss's boss?
     See "Navy Organization"
- What do all those badges and things mean?
     See "Reading Uniforms"
- Why do I care what OPNAV NOTICE 5400 is?
     See "The Paper Navy"
- What do I do when the bugles start sounding?
     See "Navy Customs and Traditions"
- Why does the chief across the hall always wince when I use the word "boat"?
     See "Ships"
- How do I keep from getting lost if I go aboard an aircraft carrier?
     See "Shipboard"
- What in the world is an F/A-18E/F?
     See "Aircraft"
- How do I decipher such things as AN/SPY-1D?
     See "Weapons"
- Just when *is* 1337!?!
     See "Lucky Bag"
- What do NAVSEA and LANTPATRAMID mean?
     See appendix B

I have chosen an approach that I hope will make this book more useful to busy people. The main part of the chapter will provide all the information you will need (or tell you where to find more). But for quick review and reference purposes, I have included a "QuickRef" section that summarizes the main points detailed in the preceding chapter. With this approach, you can get the essentials from the QuickRefs and a more detailed explanation from the main part of the chapter.

A final note before we begin. I love the Navy. It has been more things to me than I could possibly explain. But just as loving parents can be critical of their child, I will on occasion make remarks in these pages that may sound critical or sarcastic. These are simple recognitions of the obvious—that the Navy has some strange, sometimes totally illogical characteristics born of age-old traditions and evolutionary compromises that can make understanding difficult (see the discussion of warrant officers in chapter 1, "Military Titles," for example). Though these things are the by-product of a rich heritage, for me to not acknowledge them as confusing and even, on occasion, seemingly absurd (and thereby sound critical) might well shake your confidence by making you wonder if one of us is losing his mind. So take these occasional judgmental remarks as they are intended: critical perhaps, but no less loving at the same time.

Now, turn the page and enter the nautical, military, governmental, traditional, futuristic, confusing, fascinating, _important_ world of the Navy. Welcome aboard!

NavCivGuide

CHAPTER ONE

Military Titles

This chapter will help you understand the many titles that military people have, and how to recognize them in the various alphanumeric designations that military personnel use to identify and distinguish themselves. You will learn the rank structures, what the differences are in such things as paygrades and ratings, and, more practically, how to properly address military people, both in person and in writing.

PAYGRADES, RANKS, RATES, AND RATINGS IN THE NAVY

One of the things most alien to the newcomer to the Navy is its hierarchy and all of the many titles that military people have. These titles fall into a number of categories, such as ranks and ratings, and understanding the distinctions among them can go a long way toward understanding the Navy as a whole.

To begin with, it is important to distinguish between a "billet" (or "current assignment") and a rank. A billet is much like a job title elsewhere in the world. Just as the head of a corporation might be called a "chief executive officer," so a military person might be called a "chief" (as in "Chief of Naval Operations") or a "commander" (as in "Commander 7th Fleet"). Some examples of other billet titles (among thousands) are "Work Center Supervisor," "Combat Systems Officer," "Leading Seaman," "Executive Officer," "Deck Division Chief," "Ship's Secretary," and "G Division Officer."

One thing that distinguishes the military from most of corporate America is that military people also have ranks. Those who have worked in the federal government as civilians will find the idea of ranks a little less alien because of the "GS" (General Schedule) system that gives a government employee a GS rating, establishes what that person will be paid, and places them within a hierarchy of relative authority and responsibility. Ranks in the military are similar but a bit more complex. For one thing, people in the government have to ask one another what their GS ratings are, whereas military personnel wear their ranks on their sleeves, collars, or shoulders. But, like GS ratings, ranks denote a person's ability to take on responsibility and authority, and they also determine paygrades. In fact, the one thing that military personnel from different services have in common with each other is their paygrade. A man enlisting in the Air Force and a woman enlisting in the Navy will both have a paygrade of "E-1" even though he will have the rank of "Airman Basic" and she will be a "Seaman Recruit."

All of this is to say that a newcomer to the Navy should be aware that billets and ranks are related (a person might have to have the *rank* of lieutenant in order to be eligible to fill a specific *billet*, such as being the weapons officer on a particular kind of ship) but the two are also separate in their own ways.

There are thousands and thousands of billets in the Navy just as there are many job titles in any large company. But there are a much smaller number of ranks, and you will go a long way toward understanding what the Navy is all about by learning a bit about the Navy's (and the other services') ranks.

Before moving on to rank titles, one word of clarification about billets in the Navy. The heads of many units (such as ships, aircraft squadrons, etc.) are known by the generic billet title of "commanding officer." This is true of all the armed services. But, with a nod to tradition, the Navy *also* uses the term "captain" for many of these billets (as in, "He is the captain of that destroyer," or, "She is the captain of that cruiser"). But you will soon see that "captain" is also the name of a rank in the Navy (and in the other armed services as well—though at a different level). This means that the commanding officer of a destroyer might hold the rank of commander but still be called the "captain" of that ship. And the captain of an aircraft carrier usually holds the *rank* of captain as well.

A word of warning and encouragement. As you can see in the previous paragraph, understanding titles in the military—be they billet titles or rank titles—can be very confusing and therefore somewhat daunting. But keep in mind that seventeen-year-olds come into the Navy every day with no former knowledge, and they learn it—quickly and well. You can handle it!

Perhaps the best way to begin understanding the rank structure of the Navy is to look at table 1.1. After your first reaction of "You—have—got—

Table 1.1. Navy Paygrades and Ranks

Paygrade	Ranks (and rates)
O-10	Admiral
O-9	Vice Admiral
O-8	Rear Admiral (Upper Half)
O-7	Rear Admiral (Lower Half)
O-6	Captain
O-5	Commander
O-4	Lieutenant Commander
O-3	Lieutenant
O-2	Lieutenant (Junior Grade)
O-1	Ensign
W-5	Chief Warrant Officer
W-4	Chief Warrant Officer
W-3	Chief Warrant Officer
W-2	Chief Warrant Officer
W-1	Warrant Officer (Not currently in use.)
E-9	Master Chief Petty Officer of the Navy
"	Fleet Master Chief Petty Officer
"	Force Master Chief Petty Officer
"	CNO-Directed Command Master Chief Petty Officer
"	Command Master Chief Petty Officer
"	Master Chief Petty Officer
E-8	Senior Chief Petty Officer
E-7	Chief Petty Officer
E-6	Petty Officer First Class
E-5	Petty Officer Second Class
E-4	Petty Officer Third Class
E-3	Seaman (or Airman, Fireman, Constructionman, Hospitalman)
E-2	Seaman Apprentice (or Airman Apprentice, Fireman Apprentice, etc.)
E-1	Seaman Recruit (or Airman Recruit, Fireman Recruit, etc.)

to—be—kidding!" we can begin the process of making some sense out of it. (Remember those seventeen-year-olds!)

Paygrades

To begin with, let's look at the column marked "Paygrades." As stated before, these are common to all the armed forces and are what determine the basic pay of a person in the military. There are E-1s through O-10s in all the armed services, though they will be called different things as we shall soon see. A young man entering the Navy will be paid as an E-1 and will be called a "Seaman Recruit" until he successfully completes Recruit Training (commonly called "Boot Camp"). His first promotion will be to E-2, at which time he will receive a pay raise and will be then be a "Seaman Apprentice" until his next promotion.

Note that there are some alternatives in title that depend upon what occupational part of the Navy the young person is slated (by desire and qualifications) to serve in. For example, a young woman who enters the Navy with a follow-on assignment after Boot Camp to attend a school in shipboard engineering and then to serve in a ship as a gas turbine technician would become a "Fireman Recruit" (instead of Seaman Recruit) upon entry, and her first promotion will be to Fireman Apprentice. Those who will be working in aviation occupational specialties will be "Airman Recruits"; those who are slated to work in construction related occupations (known popularly as "SeaBees," or "CBs," short for Construction Battalion) will be "Constructionman Recruits"; and those who will be working in medical or dental occupations will be "Hospitalman Recruits."

Officers and Enlisted

Looking at table 1.1, it might at first seem obvious that Sailors would start at the bottom (E-1) upon entering the Navy and move up through the various levels until they either reach the top or leave the Navy for another career or retire. But of course it couldn't be that simple! Note that the table does not go from E-1 at the bottom to E-something at the top. Instead, it shifts from "E" to "W" and "O" scales along the way. This is because the Navy (like all the services) has enlisted Sailors and officer Sailors (with warrant officers in between). In earlier times—before the United States of America changed the world with its successful democracy—the militaries of Europe differentiated between officers and enlisted based upon social class. If you were of so-called noble birth and entered military service, you would be an officer, and as a result of good performance (or too often because of whom you knew), you could aspire to reach the levels of command and perhaps go beyond to become a general or an admiral. If you were of more common birth, your only choice was to enter the army or navy as a foot soldier or deck hand, and though you could be promoted, there was a glass ceiling you could never penetrate because of your social class.

EDUCATION

Even though our Army and Navy were modeled after the armies and navies of Europe, this class system was obviously not going to work in a democratic America. Various means of emulating yet changing this system were tried—including the election of officers—but what eventually evolved was a system based primarily upon education. Although not quite this neat and simple, a reasonable way to look at the system that evolved (and which is still basically in effect today) is to think of officers as those individuals who enter the service with college degrees already completed and enlisted as those who enter the service without a degree. There are numerous exceptions and variations to this "rule" and those lines are blurring today for a variety of reasons, but it is still a reasonably accurate way to understand the system. Another analogy that is not entirely accurate but may be helpful in understanding the differences is to think of enlisted and officers as roughly equivalent to labor and management respectively.

With the above in mind, you can see that a young man fresh out of high school who decides that he wants to serve in the Navy and work on airplanes would enter the service as an Airman Recruit with a paygrade of E-1. After completing Boot Camp, he would be promoted to Airman Apprentice (E-2) and subsequently move up through the enlisted paygrades (E-3, E-4, and so on). A young woman fresh out of college on the other hand would more likely enter the Navy as an officer, beginning her service as an Ensign with a paygrade of O-1. She could then move up through the officer ranks as a Lieutenant (Junior Grade) (O-2), then a Lieutenant (O-3), and so on.

In a "normal" career, the young man who enlisted in the Navy could aspire to make Master Chief Petty Officer (E-9) in a very successful career. The young woman could reasonably hope to become an Admiral in her very successful career.

EXCEPTIONS

Keep in mind that there are many exceptions to this simple pattern I have described. One exception is that a person may enter the service with a college degree but may prefer to be enlisted rather than become an officer. Another exception is that some young men and women who have demonstrated the appropriate potential may receive appointments to the U.S. Naval Academy, in which case they will enter the service without a college degree but will earn one at the Academy and become an O-1 upon graduation. There are also many ways that enlisted Sailors can become officers part way through their careers. One example is the Seaman to Admiral Program, which selects enlisted Sailors (with the right qualifications and desire) to earn a college degree at the Navy's expense and then become an officer upon graduation.

WARRANT OFFICERS

Yet another exception is for enlisted Sailors to be recognized as so proficient in their Navy occupations that they are promoted to "Warrant Officer." These are indicated as W-1 through W-5 paygrades in table 1.1. One would think that an individual selected to be a warrant officer would become a W-1 on the paygrade chart in table 1.1, but such is not the case. For reasons too complicated to explain here, the Navy no longer uses the W-1 paygrade; Sailors selected to be warrant officers go directly to W-2 (Chief Warrant Officer) and then move up to W-3, and so on. This is the sort of thing I was referring to in the "loving parent" speech I made earlier. When the Navy decided to go from three to four warrant officer ranks, it did not shift back to using the W-1 rank, but instead created a W-5 rank!

TERMINOLOGY

People who enlist in the Navy are generically called "enlistees" or "enlisted personnel" and serve specifically contracted periods of time called "enlistments."

People who enter the Navy as officers (or later become officers) are referred to generically as "officers" and are said to be "commissioned." Their commissions come from the President of the United States and are open-ended in time, ending only when the officer resigns, or is retired, or is dismissed from the service. Although officers do not sign on for specific enlistments as enlistees do, they do sometimes incur periods of obligated service—as "payback" for going to the Naval Academy or flight school, for example—that prevent them from resigning before that obligation is met.

People who are selected to become warrant officers from the enlisted ranks are generically called "warrants." When there was a W-1 paygrade, those individuals were said to receive "warrants" from the Secretary of the Navy, but the existing W-2, W-3, W-4, and W-5 ranks all receive commissions from the President as do the officers with paygrades O-1 through O-10.

Although there are different terms used to distinguish officers and enlisted, all people serving in the Navy on active duty or in the Navy Reserve are known as "Sailors."* You can never go wrong calling anyone in a Navy uni-

---

*The term is capitalized, just as "Marine" is, by direction of the Secretary of the Navy.

form a "Sailor." This was not always the case. In the past, the term "sailor" was often used to describe only enlisted people. But in more recent times, "Sailor" now applies to all Navy personnel in uniform—although you may encounter a "dinosaur" who still makes the old distinction.

One holdover remains, although it may eventually go away: when making a distinction between enlisted and officer personnel, the term "enlisted Sailor" is often used, but "officer Sailor" is not usually used. So you may encounter something like, "Many enlisted Sailors were there, but not many officers attended the seminar."

Ranks, Rates, Ratings, and Other Titles

Now for some more confusion. Though the Army, Air Force, and Marine Corps use ranks as one would expect, the Navy (and Coast Guard) also use the terms "rate" and "rating" in distinctive ways. Rates are very similar (if not identical) to ranks and are the titles used to describe the various levels of pay-grade for enlisted Sailors. Ratings are titles that apply to enlisted occupations.

Rate

We have already defined "ranks" as specific titles linked directly to paygrade and representing degrees of authority and responsibility. In the discussion above, we used the term "rank" to describe the various levels illustrated in table 1.1. But you should be aware that in the Navy (and the Coast Guard) there is another distinction sometimes (but not always) made. Adhering strictly to tradition, officers have ranks, whereas enlisted people have "rates." This may be a dying tradition that will not be missed once it reaches its final demise, because it serves no useful purpose and is often confusing. But you will still encounter it, so it is helpful to understand it.

Rating

The use of the term "rate" becomes even more confusing when we consider that there is yet another term used in the Navy (and the Coast Guard)—this one with a more distinctive and useful purpose. "Rating" (as opposed to "rate") is the term used to describe an enlisted Sailor's specialty or occupation in the Navy, based upon knowledge and skills. In the other services, these specialties are commonly called MOS (for "Military Occupational Specialty").

These ratings have specific names—like "Gunner's Mate" or "Sonarman" or "Yeoman"—and each has a specific symbol that is (unlike MOSs in the other services) actually worn on the uniform as part of the "rating badge" (see chapter 3). It is beyond the scope of this book to go into great detail regarding ratings, but the Navy's current ratings, with their symbols, are included in figure 3.4 in chapter 3 if you are interested in learning more.

There are numerous ratings in the Navy. The number varies as occupational needs change; for example, there once was a "Sailmaker" rating, but that has been discontinued for obvious reasons. Most are obvious by their titles, such as "Intelligence Specialist," "Electronics Technician," and "Musician." But others—like "Quartermaster" (navigation specialist)—are less so. Some can be misleading, such as "Fire Controlman," which one might surmise is a firefighter but is actually one who controls the firing of guns or missiles (firefighting specialists in the Navy are "Damage Controlmen").

Rates change as a person gets promoted, but ratings carry over with each promotion. A memory aid that may help you to remember which is which of the two similar terms of rate and rating is to realize that rate and rank both have four letters.

Tying It All Together

Officer ranks are pretty straightforward. A person is an "Ensign" with a pay-grade of O-1 or a "Commander" (O-5), and so on. But enlisted titles are a bit more complex.

As already explained, at the E-1 through E-3 levels, a Sailor can be a Seaman Recruit (E-1), a Seaman Apprentice (E-2), or a Seaman (E-3) if he or she enters the Navy with the expectation of serving in various occupations relating to ships. But if he or she is going to be working with aircraft, E-1 through E-3 become Airman Recruit, and so on, instead of Seaman. Other options include Fireman (engineering occupations), Constructionman, and Hospitalman.

At the E-4 through E-9 rates, it gets a little more complicated. Once Sailors are promoted to E-4, they become "rated," meaning that have a specific rating (that is, an occupation such as Gunner's Mate). At that point they actually have *two* titles: Petty Officer Third Class (see table 1.1) and Gunner's Mate Third Class. They will usually remain in that rating for the rest of their careers, promoting to Petty Officer Second Class (E-5) and simultaneously to Gunner's Mate Second Class, then to E-6 (Petty Officer First Class and Gunner's Mate Second Class), E-7 (Chief Petty Officer and Chief Gunner's Mate), and so on. These titles are pretty much interchangeable. The good news is that you do not need to know all of these titles to get by. There are many ratings, and few people have them all memorized. Just knowing the levels of petty officer and recognizing that ratings may be used in their place is sufficient. How to do this is explained in more detail below.

One word about usage. You can reverse the terminology, depending upon the circumstances. For example, if you know someone who is an E-5, you would speak or refer to them (formally) as "Petty Officer Second Class Jones," but when referring to them without their name, it is also appropriate to call them a "second class petty officer" (in other words putting the "second class" before the "petty officer"). For example, in conversation you might say, "This is Petty Officer First Class John Jones. I first met him when he was a third class petty officer at the Pentagon." This is a small point but worth knowing to avoid confusion.

Keep in mind that chief petty officers are always referred to in that order, never "Petty Officer Chief." But you may (particularly when in less formal circumstances) drop the "petty officer" part. For example, the following is correct: "This is Chief Petty Officer Jane Jones. Chief Jones and Senior Chief Smith were shipmates together in USS *Independence* back before they were promoted to chief."

Warfare Specialties

Another "title" common to many Sailors of various ranks, ratings, and so on is their warfare specialty, obtained through an additional qualification process (involving schooling, practical experience, or a combination of both). Both officer and enlisted Sailors may have one or more of these specialties. These are in addition to, and independent of, their rank, rate, and rating. Some— but not all—of these specialties are surface warfare, air warfare, submarine warfare, and special warfare. Enlisted Sailors append identifying letters in parentheses after their rates (for example, "BMC (SW) John P. Jones USN"); some examples of identifying letters are "SW" for surface warfare, "AW" for air warfare, and "SS" for submarine warfare. Officers do not append their warfare specialties to their ranks, but *staff* officers do append them after their names (see below).

For the most part, you can ignore these in routine communications, but it does not hurt to be aware of them.

Additional Officer Titles

As mentioned above, officers do not append alphabetic representations of their warfare specialties to their names as enlisted personnel do. But "staff officers" do add letters to indicate their specialties.

Officers can be roughly divided into two groups: line and staff. The term "line officer" stems from the days of sail when groups of ships would form lines of battle to face each other in combat. The officers who commanded (or aspired to command) these ships were referred to as line officers. Staff officers were those who were less likely to be on the line of battle but would instead support the Fleet in other important ways.

Today, generally speaking, line officers are those who are eligible for command at sea; they command or may someday command ships, submarines, aircraft squadrons, and so on. Staff officers fill important support functions as doctors, lawyers, chaplains, supply corps officers, and more. These staff officers often append letter combinations (such as "SC" for Supply Corps, "CHC" for Chaplain Corps, "JAG" for Judge Advocate General [lawyer], etc.) to their names (for example, "LT Clarence Darrow, JAG, USN").

As with the abbreviations for enlisted warfare specialties, these can be ignored in routine communications but are nice to know on occasion.

MILITARY ALPHABET SOUP

Having gained some understanding of the different kinds of titles used in the Navy, it is now time to crack the code on all the cryptic abbreviations that go along with those titles. I call them "alphabet soup," but in truth "alphanumeric soup" would be a better description.

By the time we are finished, such things as "BM2 (AW, SS, SW) John P. Jones USN" and "RDML Wendy T. Door, CHC, USN" will make a lot more sense. Understanding these things will permit you to communicate more effectively in the Navy and will ease that unsettling feeling of being somehow excluded from the "secret society." Some of the ability to decipher these alphanumeric "codes" will rely upon simple understanding and a bit of memorization, but we will also come up with some shortcuts and hints that will simplify the task.

The first thing to realize is that for most purposes you need only concentrate on the letters and numbers preceding a Sailor's name. Those other letters in parentheses following the person's name are specialties and you can ignore them for basic communications. The last group of letters (USN, USCG, etc.) merely identify a person's service (Navy, Coast Guard, etc.). Just keep in mind that when addressing military personnel, *it's what's up front that counts most*.

Addressing the Navy

There are "rules" for addressing Sailors in the Navy and, unfortunately, they vary somewhat with circumstances. Let's begin by expanding table 1.1 a bit and calling it table 1.2. Keeping this table handy and referring to it for a while will eventually lead to your using these forms of address as second nature. As with table 1.1, we see the paygrades (E-1 through E-9, W-1 through W-5, and O-1 through O-10) and the names of the ranks (also known as "rates" for enlisted Sailors). But this table has some additional useful features.

The "Abbreviations" column lists the proper abbreviations of the ranks *as they are used by the Navy*. Be aware that you will see other abbreviations used elsewhere; civilian book publishers do not like to use all capital letters because they feel it is distracting on a printed page, and though the Navy's abbreviations have a certain military logic to them, they are not always clear to people not familiar with the Navy. As a result, you will often see such things as "Lt. Cmdr." (instead of LCDR), or "Vice Adm." (instead of VADM), or "Smn" (instead of SN). But when dealing with or within the Navy, you will have more credibility if you use the Navy's versions of abbreviations as shown in table 1.2. These abbreviations are used fairly often in the Navy on official forms or in documents. They are also used on the outside envelope when addressing letters to people in the Navy.

If you were going to send a letter to Seaman Apprentice Grace Hopper, you would write her name on the outside of the envelope as "SA Grace Hopper USN." Inside the letter, you should again use the Navy abbreviation in the address part of the letter, but in the salutation, you should use the spelled-out form provided in table 1.2's column labeled "Dear—." See the example below:

30 August 2007

SA Grace Hopper USN
USS Ronald Reagan (CVN 76)
FPO AP 96616-2876

Dear Seaman Hopper:

Your copy of *The Bluejacket's Manual* was found in this office and is enclosed.

Sincerely,

J. Q. Adams

John Q. Adams

Table 1.2. Navy Paygrades, Ranks, Abbreviations, Salutations, and Forms of Address

Paygrade	Ranks (and rates)	Abbreviations[a]	Dear—[b]	Direct address[c]
O-11	Fleet Admiral	FADM	Not currently in use.	
O-10	Admiral	ADM	Admiral	Admiral
O-9	Vice Admiral	VADM	Admiral	Admiral
O-8	Rear Admiral (Upper Half)	RADM	Admiral	Admiral
O-7	Rear Admiral (Lower Half)	RDML	Admiral	Admiral
O-6	Captain	CAPT	Captain	Captain
O-5	Commander	CDR	Commander	Commander
O-4	Lieutenant Commander	LCDR	Commander	Commander
O-3	Lieutenant	LT	Lieutenant	Lieutenant
O-2	Lieutenant (Junior Grade)	LTJG	Lieutenant	Lieutenant
O-1	Ensign	ENS	Ensign	Ensign
W-5	Chief Warrant Officer	CWO5	Chief Warrant Officer	Warrants are usually addressed
W-4	Chief Warrant Officer	CWO4	Chief Warrant Officer	by their specialty; as in "Boatswain,"
W-3	Chief Warrant Officer	CWO3	Chief Warrant Officer	"Gunner," etc.
W-2	Chief Warrant Officer	CWO2	Chief Warrant Officer	
W-1	Warrant Officer	Not currently in use.		
E-9	Master Chief Petty Officer of the Navy	MCPON	MCPON	MCPON (pronounced "mick-pon")
"	Fleet Master Chief Petty Officer	FLTCM	Fleet Master Chief	Fleet
"	Force Master Chief Petty Officer	FORCM	Force Master Chief	Force
"	CNO-Directed Command Master Chief Petty Officer	CNOCM	CNO-Directed Command Master Chief	Master Chief
"	Command Master Chief Petty Officer	CMDCM	Command Master Chief	Master Chief
"	Master Chief Petty Officer	MCPO[d]	Master Chief	Master Chief
E-8	Senior Chief Petty Officer	SCPO[d]	Senior Chief	Senior
E-7	Chief Petty Officer	CPO[d]	Chief	Chief

Table 1.2. (*continued*)

Paygrade	Ranks (and rates)	Abbreviations[a]	Dear—[b]	Direct address[c]
E-6	Petty Officer First Class	PO1[d]	Petty Officer	Petty Officer
E-5	Petty Officer Second Class	PO2[d]	Petty Officer	Petty Officer
E-4	Petty Officer Third Class	PO3[d]	Petty Officer	Petty Officer
E-3	Seaman (or Airman, Fireman, Constructionman, Hospitalman)	SN (or AN, FN, CN, HN)	Seaman (or Airman, Fireman, Constructionman, Hospitalman)	Seaman (or Airman, Fireman, Constructionman, Hospitalman)
E-2	Seaman Apprentice (or Airman Apprentice, Fireman Apprentice, etc.)	SA (or AA, FA, CA, HA)	Seaman (or Airman, Fireman, Constructionman, Hospitalman)	Seaman (or Airman, Fireman, Constructionman, Hospitalman)
E-1	Seaman Recruit (or Airman Recruit, Fireman Recruit, etc.)	SA (or AA, FA, CA, HA)	Seaman (or Airman, Fireman, Constructionman, Hospitalman)	Seaman (or Airman, Fireman, Constructionman, Hospitalman)

[a] Use the "Abbreviation" column for addressees on letters and envelopes (e.g., "LCDR Stephen Decatur USN").
[b] Use the "Dear" column (followed by person's surname) for salutations and introductions (e.g., "Dear Commander Decatur").
[c] Use the "Address" column when addressing someone directly (e.g., "Good morning, Commander," or, "Force, the admiral wants to see you on the flag bridge").
[d] Rating abbreviations (such as BMC or QM2, etc.) are used more often than the rate abbreviations CPO, PO2, and so on.

The last column in table 1.2 ("Direct Address") tells you how you should address a Sailor in polite conversation, as in, "Good morning, Captain Halsey," or, "Chief Gatlin, would you please explain why you have all those stripes on your lower left sleeve?"

When introducing a Sailor to someone else, you should use his or her full title the first time but then use the "direct address" form from then on. For example: "Doctor Thomas Dooley, this is Lieutenant Commander Stephen Decatur. Commander Decatur is headed for the Mediterranean next month."

Note that these addresses are mostly logical and simple, but there are a few quirks. In most cases, brevity rules. Lieutenant Commander becomes simply "Commander," "Seaman Apprentice" becomes "Seaman," and so on. Note, however, that you should not address a Senior Chief Petty Officer as "Chief" and that "Master Chief Petty Officer" does *not* become "Master."

There are a few other "quirks" you should be aware of. There is only one Master Chief Petty Officer of the Navy, and he or she is the most senior enlisted Sailor in the Navy. More by convention than anything formal, he or she is referred to as "the MCPON" and is often addressed as "MCPON." As indicated in the table, this is pronounced "mick-pon" ("pon" rhyming with "john," not "loan").

Note that there are multiple forms of E-9 and that their addresses require a little extra attention. This is a case where billets and ranks overlap a bit. A master chief petty officer who is assigned as the principal enlisted advisor to the commanding officer of a ship or some other unit would take on the title of "Command Master Chief Petty Officer" (more frequently referred to as simply "the Command Master Chief"). A master chief who has been assigned as the principal enlisted advisor to a fleet commander would be the "Fleet Master Chief." People filling these billets not only take on the new rank names as indicated, but they also wear different rating badges on their left sleeves.

Warrant officers present another challenge. They are often called by their specialty as indicated in table 1.2, and you will probably have to learn these on a case by-case basis.

I have included the rank of "Fleet Admiral" in table 1.2 because it technically exists and was once used. In World War II, Admiral Leahy, Admiral King, Admiral Nimitz, and Admiral Halsey were promoted to Fleet Admiral (also known as "five-star Admiral") but no one has been given that rank since, and it is doubtful it will ever be used again.

Note that a newly promoted admiral becomes a "Rear Admiral (Lower Half)" and, if later promoted, would become a "Rear Admiral (Upper Half)." These odd titles are compromises reached after years of trying different titles. An O-7 was a "Commodore" during World War II; at one time, O-7s and O-8s both had the title of "Rear Admiral" with no distinctions in title; and, for a while, an O-7 was called a "Commodore Admiral." The titles in table 1.2 are the currently correct ones. A good-natured rear admiral (lower half) who had just received word that he had been promoted to rear admiral (upper half) quipped, "That's great news. Now I can wear my shoulder boards on my shoulders instead of my hips!"

One last clarification. As noted above, "Commodore" was a rank in the Navy for a time. Though that has changed, you may still encounter the term; but if you do, be aware that it is probably associated with a billet rather than a rank. This comes about when an officer who is not an admiral is given command of a group of ships (such as a destroyer squadron). Each ship has its own commanding officer (also known as "Captain" as explained earlier), so to avoid confusion (and to grant a degree of honor), the officer in charge of the group is called the "Commodore." She or he may hold the rank of "captain" but is referred to as "Commodore" by virtue of commanding more than one ship.

Cracking the "Codes"

Making sense of all that alphabet soup surrounding a Sailor's name is a bit daunting, but there are some ways to make it easier. Remember to concentrate on the letters and numbers preceding the person's name to keep things simple.

As mentioned above, officers are pretty straightforward. You need only familiarize yourself with the ten ranks in use and you will be able to address them as indicated in table 1.2.

Enlisted Sailors seem much more complicated, but there are methods you can use and clues you can look for that will help simplify it all for you.

Petty Officers

A complicating factor is this coexistence of rates and ratings, that an enlisted Sailor can be a Petty Officer Third Class and a Gunner's Mate Third Class at the same time. In practice, you will rarely see the abbreviations listed in table 1.2 used for paygrades E-4 through E-9; the common practice is for Sailors to use their rating abbreviation instead. The odd thing is that even though Sailors use their ratings most of the time, *you do not need to* when addressing or responding to them. Using the clues explained below, you may convert their ratings to the more simple rate.

Let's focus on a young third class petty officer and call him "Christopher Garrett." We know from tables 1.1 and 1.2 that a Sailor who has attained the paygrade of E-4 is a Petty Officer Third Class. To be a third class, he must also have a specialty occupation known as a rating. If he is a sharp Sailor, he may have already attained his warfare qualification in surface warfare. All of that means that he would sign an official log book or appear on a watch bill as

GM3 (SW) Christopher Garrett USN

Now for the good news. In addressing him, you can look at all that alphabet soup and know that simply calling him "Petty Officer Garrett" is all you need to do.

The warfare qualification ("SW") always appears in parentheses after the name and can be ignored for all intents and purposes. Though it is important to young Garrett and something he is rightfully proud of, it has no bearing on how you communicate with him.

The "USN" after his name simply indicates that he is in the Navy as opposed to one of the other services (all of which are logical and easily deciphered: USA for Army; USMC for Marine Corps; USCG for Coast Guard; and USAF for Air Force).

So the only tricky part is the "GM3" before his name. And if you understand that the "GM" signifies his rating, all you need focus on is the "3" to know that he is a Petty Officer Third Class. That is all you really need to know to be able to properly address him. The fact that "GM" stands for "Gunner's Mate" is nice to know—and certainly important to him—but it is not necessary knowledge for you to be able answer a letter from him or to greet him properly in person.

What does all this mean? Simply that *the "3" is really the only clue you need.*

If he sent you a letter a year later and identified himself as "GM2 (SW, AW) Christopher Garrett USN," you would repeat all the alphabet soup on the outside of the envelope and in the address part of your response to him, but you still need only focus on the "2" to know that he should be addressed as "Dear Petty Officer Garrett" in the salutation.

And if a young woman sends you a letter and signs it "QM1 (SW) Rachel Alexander USN," all you would need to focus on is the numeral "1" to know that you should answer her letter "Dear Petty Officer Alexander."

Bottom line: if you see a 1, 2, or 3 in a person's alphabet soup, you should address him or her simply as "Petty Officer (name)."

It may help to know that *almost* all ratings are signified by two letters, so you would expect to see two letters before the numeral in a person's name. Unfortunately, there are a few ratings that are signified by three letters, rather than merely two, so don't let it throw you if you see something like "ABF3 Reuben James USN." You do not need to know that he is an "Aviation Boatswain's Mate (Fuels)," only that he is a petty officer.

Chief Petty Officers

Having mastered E-4 through E-6 in table 1.2, we can move up to E-7 through E-9. As a few more years pass and Christopher Garrett continues to be promoted, he reaches E-7, and referring to table 1.2, we see that he is now a "Chief Petty Officer" and his alphabet soup now becomes

GMC (SW, AW) Christopher Garrett USN

As before, his rating is virtually irrelevant for our purposes and the pertinent clue is found following those initial letters (GM). But now we see that he has traded his numbers for a "C" to symbolize his new position and title. Now we must be sure to recognize his achievement by calling him "Chief Petty Officer Garrett" or, more often, simply "Chief Garrett."

Bottom line, if you see a "C" at the end of the initial part of a Sailor's alphabet soup, you know that he or she should be addressed as "Chief."

Once Chief Garrett is promoted to E-8, we know from table 1.2 that he is now a Senior Chief Petty Officer. If you think that the logical thing to do would be to add an "S," you would be correct, but beware: the "S" is appended *after* the "C," not before. So now, Garrett's alphabet soup becomes

GMCS (SW, AW) Christopher Garrett USN

So the bottom line here is that if you see the letters "CS" at the end of the initial portion of alphabet soup, you know that individual is an E-8 and is properly addressed as "Senior Chief Garrett."

Promotion to E-9 makes a person a "Master Chief Petty Officer." The "logic" applied to senior chiefs is continued to master chiefs. Senior Chief Garrett becomes

GMCM (SW, AW) Christopher Garrett USN

The "M" indicates master and follows the "C." Bottom line is that when you see the letters "CM" at the end of the initial portion of alphabet soup, you know that individual is an E-9 and is properly addressed as "Master Chief Garrett."

As explained earlier, there are some additional E-9 titles in table 1.2. These are given to some exceptional individuals who have been chosen to hold very important positions in the Navy. Note that there are a few conventions in addressing these individuals that you should be aware of (such as directly addressing a Fleet Master Chief Petty Officer as "Fleet.")

Nonrated Sailors

Having explained E-4 through E-9, we need only cover E-1 through E-3 to complete our quest. In some ways these are the toughest, because there are a number of variations. We have already seen that a Sailor enters the Navy as either a Seaman, a Fireman, an Airman, a constructionman, or a Hospitalman, depending upon what rating they are expected to pursue. In other words, when Christopher Garrett entered the Navy to become a Gunner's Mate, he came in as a Seaman Recruit (because Gunner's Mates fall into the Seaman group—along with Boatswain's Mates, Yeomen, and many others) and then became a Seaman Apprentice and eventually a Seaman before becoming a Gunner's Mate at the E-4 level. If Rachel Alexander comes into the Navy and her desire and aptitude make her a good candidate to eventually enter an aviation rating, such as Aviation Electrician's Mate, she will be designated an "Airman Recruit" as an E-1 and will promote to "Airman Apprentice" and "Airman" (E-2 and E-3, respectively).

Sailors become "rated" (i.e., take on a specific occupational specialty known as a "rating") at the E-4 level. They cannot become an E-4 (Petty Officer Third Class) without also becoming a specific rating at the same time. Conversely, Sailors at the E-1, E-2, and E-3 levels are considered (and often referred to as) "nonrated."

Now one last bit of confusion, and then some simplification. Nonrated Sailors who achieve certain milestones—such as successful completion of a school in their chosen rating—are known as "strikers." In Navy parlance, they are "striking for" a certain rate. If they are formally designated as a striker, they append the appropriate letters (of the rating) to the ones for their rate. For example, let's say that young Garrett graduates from Boot Camp and is promoted to E-2 (Seaman Apprentice). His alphabet soup at that point would be "SA Christopher Garrett USN." After Boot Camp, he goes to what the Navy calls a "Class A School" to learn the skills of a Gunner's Mate. Upon successful completion of the school, he is formally designated a Gunner's Mate striker. At that point his "soup" changes to "GMSA Christopher Garrett USN." When next promoted, he will become GMSN Christopher Garrett USN. He will be still be considered "nonrated" but will also be recognized as a striker.

Bottom line is that just as with the petty officers and chief petty officers, all you really need to look for are some key letters. If you see an "SA" or an "SN," either standing alone or after some other letters, you know that Sailor should be addressed as "Seaman." Someone with an "FA" or "FN" should be addressed as "Fireman." The appropriate address for "AA" and "AN" is "Airman." "CA" and "CN" are called "Constructionman," and "HA" and "HN" are "Hospitalman."

So if you encountered "AEAN Rachel Alexander USN," you would know that you should call her "Airman Alexander," because of the letters "AN" (you may ignore the letters "AE," which signify that she is an Aviation Electrician's Mate striker). Once she is promoted to E-4, she will become an Aviation Electrician's Mate Third Class (AE3 Rachel Alexander USN), and remembering the clues explained earlier, you would now call her simply "Petty Officer Alexander."

Summary

All of the above can be boiled down to some relatively simple rules. Focus on the part of the alphabet soup preceding a person's name and look for the following giveaways:

- Look for a number.
    1, 2, or 3 = Petty Officer
- Look for a "C" at the end.
    C = Chief
    CS = Senior Chief
    CM = Master Chief
- Look for an "A" or an "N" at the end.*
    SA or SN = Seaman
    FA or FN = Fireman
    AA or AN = Airman
    CA or CN = Constructionman
    HA or HN = Hospitalman

Here are a few examples:

- ET2 John Sylvania USN = Petty Officer Sylvania
    He is an Electronics Technician Second Class, but you don't need to know that to address him.
- MUCM M. A. Donna USN = Master Chief Donna
    She is a Master Chief Musician, but all you need to focus on is that she is a Master Chief.
- SWSA Andrew Carnegie USN = Seaman Carnegie
    He is a Steelworker (striker), but the "SA" tells you he is a Seaman Apprentice and is called "Seaman."
- JO1 Walter Cronkite = Petty Officer Cronkite
    The fact that he is a First Class Journalist is nice to know, but "Petty Officer" is all you need.

---

*Recruits are identified by the abbreviations SR (for Seaman Recruit), FR (for Fireman Recruit), and so on, but you are unlikely to encounter these abbreviations unless you work at Recruit Training Command.

- QMC Benjamin Dutton USN = Chief Dutton
  He is a Quartermaster as well as a Chief Petty Officer, but "Chief" is all that is required.
- HMHN Clara Barton USN = Hospitalman Barton
  She is a Hospitalman striking for Hospital Corpsman, but she would only be addressed as "Hospitalman Barton."

Other Services

Just in case you were getting the hang of this, it's time to throw in some more confusion. In today's Navy, encountering, or even serving with, members of other services is not uncommon. To help, we must replicate table 1.2 for the other services.

Coast Guard

Let's begin with the Coast Guard, which is the one most similar to the Navy. In fact, looking at table 1.3, you will see that the only significant differences are that the Coast Guard only has "Seaman," "Airman," and "Fireman" (no "Constructionman" or "Hospitalman" as in the Navy), there is no W-5 pay-grade, and there are a few differences in the E-7, E-8, and E-9 ranks.

All usages—salutations, addresses, and so on—are the same as described above for the Navy. Note that the generic term for personnel serving in the Coast Guard is "Coast Guardsman," not "Sailor."

Marine Corps

The next logical stop is the other sea service, the Marine Corps. Table 1.4 reveals that there are significant differences in the Marine Corps. Note that the names of the various ranks are very different from those of the Navy and Coast Guard.

One of the more confusing aspects is that there is a rank called "Captain" in all of the services, but in the Marine Corps (and the Army and Air Force) it is at the O-3 level, and in the Navy and Coast Guard, it is at the O-6 level.

Army

Looking at table 1.5, you can see that the Army is similar, but not identical, to the Marine Corps.

Air Force

Table 1.6 reveals the similarities and differences in the Air Force.

Table 1.3. Coast Guard Paygrades, Ranks, Abbreviations, Salutations, and Forms of Address

Paygrade	Ranks (and rates)	Abbreviations[a]	Dear—[b]	Direct address[c]
O-10	Admiral	ADM	Admiral	Admiral
O-9	Vice Admiral	VADM	Admiral	Admiral
O-8	Rear Admiral (Upper Half)	RADM	Admiral	Admiral
O-7	Rear Admiral (Lower Half)	RDML	Admiral	Admiral
O-6	Captain	CAPT	Captain	Captain
O-5	Commander	CDR	Commander	Commander
O-4	Lieutenant Commander	LCDR	Commander	Commander
O-3	Lieutenant	LT	Lieutenant	Lieutenant
O-2	Lieutenant (Junior Grade)	LTJG	Lieutenant	Lieutenant
O-1	Ensign	ENS	Ensign	Ensign
W-4	Chief Warrant Officer	CWO4	Chief Warrant Officer	Warrants are usually addressed
W-3	Chief Warrant Officer	CWO3	Chief Warrant Officer	by their specialty as in "Boatswain,"
W-2	Chief Warrant Officer	CWO2	Chief Warrant Officer	"Gunner," etc.
W-1	Warrant Officer	Not currently in use.		
E-9	Master Chief Petty Officer of the Coast Guard	MCPOCG	MCPOCG	MCPOCG (pronounced "mick-pog")
"	Area Command Master Chief Petty Officer	CMC	Master Chief	Master Chief
"	Command Master Chief Petty Officer	CMC	Master Chief	Master Chief
"	Master Chief Petty Officer	MCPO[d]	Master Chief	Master Chief
E-8	Command Senior Chief Petty Officer	CSC	Senior Chief	Senior Chief
"	Senior Chief Petty Officer	SCPO[d]	Senior Chief	Senior
E-7	Command Chief Petty Officer	CC	Chief	Chief
"	Chief Petty Officer	CPO[d]	Chief	Chief
E-6	Petty Officer First Class	PO1[d]	Petty Officer	Petty Officer
E-5	Petty Officer Second Class	PO2[d]	Petty Officer	Petty Officer

Table 1.3. (*continued*)

Paygrade	Ranks (and rates)	Abbreviations[a]	Dear—[b]	Direct address[c]
E-4	Petty Officer Third Class	PO3[d]	Petty Officer	Petty Officer
E-3	Seaman (or Airman or Fireman)	SN (AN or FN)	Seaman (or Airman or Fireman)	Seaman (or Airman or Fireman)
E-2	Seaman Apprentice (or Airman Apprentice or Fireman Apprentice)	SA (AA or FA)	Seaman (or Airman or Fireman)	Seaman (or Airman or Fireman)
E-1	Seaman Recruit (or Airman Recruit, Fireman Recruit, etc.)	SR (AR or FR)	Seaman (or Airman or Fireman)	Seaman (or Airman or Fireman)

[a] Use the "Abbreviation" column for addressees on letters and envelopes (e.g., "SA Vincent Patton USCG").
[b] Use the "Dear" column (followed by person's surname) for salutations and introductions (e.g., "Dear Seaman Patton").
[c] Use the "Address" column when addressing someone directly (e.g., "Good morning, Seaman").
[d] Rating abbreviations (such as BMC or QM2, etc.) are used more often than the rate abbreviations CPO, PO2, and so on.

Table 1.4. Marine Corps Paygrades, Ranks, Abbreviations, Salutations, and Forms of Address

Paygrade	Ranks (and rates)	Abbreviations[a]	Dear—[b]	Direct address[c]
O-10	General	Gen	General	General
O-9	Lieutenant General	LtGen	General	General
O-8	Major General	MajGen	General	General
O-7	Brigadier General	BGen	General	General
O-6	Colonel	Col	Colonel	Colonel
O-5	Lieutenant Colonel	LtCol	Colonel	Colonel
O-4	Major	Maj	Major	Major
O-3	Captain	Capt	Captain	Captain
O-2	First Lieutenant	1stLt	Lieutenant	Lieutenant
O-1	Second Lieutenant	2ndLt	Lieutenant	Lieutenant
W-5	Chief Warrant Officer	CWO5	Chief Warrant Officer	Chief Warrant Officer (Gunner)
W-4	Chief Warrant Officer	CWO4	Chief Warrant Officer	Chief Warrant Officer (Gunner)
W-3	Chief Warrant Officer	CWO3	Chief Warrant Officer	Chief Warrant Officer (Gunner)
W-2	Chief Warrant Officer	CWO2	Chief Warrant Officer	Chief Warrant Officer (Gunner)

Table 1.4.   *(continued)*

Paygrade	Ranks (and rates)	Abbreviations[a]	Dear—[b]	Direct address[c]
W-1	Warrant Officer	WO	Warrant Officer	Warrant Officer (Gunner)
E-9	Sergeant Major of the Marine Corps	SgtMaj	Sergeant Major	Sergeant Major
E-9 "	Sergeant Major Master Gunnery Sergeant	SgtMaj MGySgt	Sergeant Major Master Gunnery Sergeant	Sergeant Major Master Gunnery Sergeant (Master Guns)
E-8	Master Sergeant	MSgt	Master Sergeant	Master Sergeant (Top)
"	First Sergeant	1stSgt	First Sergeant	First Sergeant
E-7	Gunnery Sergeant	GySgt	Gunnery Sergeant	Gunnery Sergeant (Gunny)
E-6	Staff Sergeant	SSgt	Staff Sergeant	Staff Sergeant
E-5	Sergeant	Sgt	Sergeant	Sergeant
E-4	Corporal	Cpl	Corporal	Corporal
E-3	Lance Corporal	LCpl	Lance Corporal	Lance Corporal
E-2	Private First Class	PFC	Private First Class	Private First Class (PFC)
E-1	Private	Pvt	Private	Private

[a] Use the "Abbreviation" column for addressees on letters and envelopes (e.g., "GySgt John Basilone USMC").

[b] Use the "Dear" column (followed by person's surname) for salutations and introductions (e.g., "Dear Gunnery Sergeant Basilone").

[c] Use the "Address" column when addressing someone directly (e.g., "Good morning, Gunny"). Addresses in parentheses are used informally.

Table 1.5.   Army Paygrades, Ranks, Abbreviations, Salutations, and Forms of Address

Paygrade	Ranks (and rates)	Abbreviations[a]	Dear—[b]	Direct address[c]
O-11	General of the Army	Not currently in use.		
O-10	General	GEN	General	General
O-9	Lieutenant General	LTG	General	General
O-8	Major General	MG	General	General
O-7	Brigadier General	BG	General	General
O-6	Colonel	COL	Colonel	Colonel
O-5	Lieutenant Colonel	LTC	Colonel	Colonel
O-4	Major	MAJ	Major	Major
O-3	Captain	CPT	Captain	Captain

Table 1.5. *(continued)*

Paygrade	Ranks (and rates)	Abbreviations[a]	Dear—[b]	Direct address[c]
O-2	First Lieutenant	1LT	Lieutenant	Lieutenant
O-1	Second Lieutenant	2LT	Lieutenant	Lieutenant
W-5	Chief Warrant Officer	CW5	Chief Warrant Officer	Mister
W-4	Chief Warrant Officer	CW4	Chief Warrant Officer	Mister
W-3	Chief Warrant Officer	CW3	Chief Warrant Officer	Mister
W-2	Chief Warrant Officer	CW2	Chief Warrant Officer	Mister
W-1	Warrant Officer	WO1	Warrant Officer	Mister
E-9	Sergeant Major of the Army	SMA	Sergeant Major	Sergeant Major
"	Command Sergeant Major	CSM	Sergeant Major	Sergeant Major
"	Sergeant Major	SGM	Sergeant Major	Sergeant Major
E-8	Master Sergeant	MSG	Master Sergeant	Master Sergeant
"	First Sergeant	1SG	First Sergeant	First Sergeant
E-7	Platoon Sergeant	PSG	Sergeant	Sergeant
"	Sergeant First Class	SFC	Sergeant	Sergeant
E-6	Staff Sergeant	SSG	Sergeant	Sergeant
E-5	Sergeant	SGT	Sergeant	Sergeant
E-4	Corporal	CPL	Corporal	Corporal
E-4	Specialist	SP4	Specialist	Specialist
E-3	Private First Class	PFC	Private	Private
E-2	Private Second Class	PV2	Private	Private
E-1	Private	PVT	Private	Private

[a] Use the "Abbreviation" column for addressees on letters and envelopes (e.g., "SGT Alvin York USA").

[b] Use the "Dear" column (followed by person's surname) for salutations and introductions (e.g., "Dear Sergeant York").

[c] Use the "Address" column when addressing someone directly (e.g., "Good morning, Sergeant").

Table 1.6. Air Force Paygrades, Ranks, Abbreviations, Salutations, and Forms of Address

Paygrade	Ranks (and rates)	Abbreviations[a]	Dear—[b]	Direct address[c]
O-11	General of the Air Force	Not currently in use.		
O-10	General	Gen	General	General
O-9	Lieutenant General	Lt Gen	General	General
O-8	Major General	Maj Gen	General	General
O-7	Brigadier General	Brig Gen	General	General
O-6	Colonel	Col	Colonel	Colonel
O-5	Lieutenant Colonel	Lt Col	Colonel	Colonel
O-4	Major	Maj	Major	Major
O-3	Captain	Capt	Captain	Captain
O-2	First Lieutenant	1st Lt	Lieutenant	Lieutenant
O-1	Second Lieutenant	2d Lt	Lieutenant	Lieutenant
W-5	No warrants in the Air Force.			
W-4				
W-3				
W-2				
W-1				
E-9	Chief Master Sergeant of the Air Force	CMSAF	Chief Master Sergeant of the Air Force	Chief
"	Chief Master Sergeant	CMSgt	Chief Master Sergeant	Chief
E-8	Senior Master Sergeant	SMSgt	Senior Master Sergeant	Sergeant
E-7	Master Sergeant	MSgt	Master Sergeant	Sergeant
E-6	Technical Sergeant	TSgt	Technical Sergeant	Sergeant
E-5	Staff Sergeant	SSgt	Staff Sergeant	Sergeant
E-4	Senior Airman	SrA	Senior Airman	Airman
E-3	Airman First Class	A1C	Airman First Class	Airman
E-2	Airman	Amn	Airman	Airman
E-1	Airman Basic	AB	Airman Basic	Airman

[a] Use the "Abbreviation" column for addressees on letters and envelopes (e.g., "Maj Gen Henry Arnold USAF").

[b] Use the "Dear" column (followed by person's surname) for salutations and introductions (e.g., "Dear General Arnold").

[c] Use the "Address" column when addressing someone directly (e.g., "Good morning, General").

SENIOR EXECUTIVE SERVICE (SES)

Military personnel are not the only ones to have a hierarchy of positions and titles. Part of the management team that makes up the leadership of the Navy is composed of a cadre of civilians officially referred to as the Senior Executive Service (SES).

The SES was established by Title IV of the Civil Service Reform Act (CSRA) of 1978 [P.L. 95-454, October 13, 1978] and became effective on 13 July 1979. The purpose was to provide a component of civilian leadership that would share the responsibility of guiding government agencies. The civilian leaders lend important perspective and an essential component to the Navy's active-reserve-civilian triad structure. The SES provides executive skills and an enhanced vision to the Navy that ensures the proper integration of civilian and military rights and responsibilities.

The CSRA's stated purpose is to "ensure that the executive management of the Government of the United States is responsive to the needs, policies, and goals of the nation and otherwise is of the highest quality." To achieve this purpose, CSRA gave greater authority to agencies to manage their executive resources and stated that the SES was to be administered to:

- attract and retain highly competent executives;
- assign executives where they will be most effective in accomplishing the agency's mission and where best use will be made of their talents;
- provide for the systemic development of managers and executives;
- hold executives accountable for individual and organizational performance;
- reward the outstanding performers and remove the poor performers; and
- provide an executive personnel system free of prohibited personnel practices and arbitrary actions.

Unlike the military, where ranks are worn on uniforms for all to see, members of the SES wear civilian clothes and are not readily identifiable. It is important that you recognize who they are, however, because they are considered the equivalent of admirals and generals and are accorded the same kinds of courtesies and protocols. One helpful clue you can watch for is the SES lapel pin, shown in figure 1.1, that many SES officials wear.

Figure 1.1   The Navy Senior Executive Service insignia.

Classifications and Appointments

There are two classifications of SES positions.

Career Reserved positions are so designated "to ensure impartiality, or public's confidence of impartiality of government." These positions can be filled only by career appointees, and they represent the majority of SES positions in the Department of the Navy.

General positions may be filled by career, noncareer, or limited appointees (see below). The General positions may be held by individuals who are advocates of the current Executive Branch policies and receive direction from the current administration. However, many General positions are held by persons with outstanding qualifications who have no political affiliation and who were brought into government to lead in a particular area for a defined length of time.

There are four types of hiring appointments that govern SES positions.

Career Appointments—Incumbents are selected by an agency merit-staffing process and must have their executive qualifications approved by a

Qualifications Review Board (QRB) convened by the Office of Personnel Management (OPM). Appointments may be to a General or Career Reserved position; rights of the individual are the same in either case.

Noncareer Appointments—Individuals are approved by OPM on a case-by-case basis, and the appointment authority reverts to OPM when the noncareer appointee leaves the position. Appointments may be made only to General positions and cannot exceed 25 percent of the agency's SES position allocation. Government-wide, only 10 percent of SES positions may be filled by noncareer appointees.

Limited-Term Appointments may be made for up to three years and are intended to expire. These appointments are nonrenewable and restricted to General positions because of the nature of the work (e.g., a special project).

Limited Emergency Appointments are also a nonrenewable appointment for up to eighteen months to a General position that has to be filled urgently.

The Federal Government, via the OPM, has about six thousand SES allocated throughout all agencies. The Department of the Navy has over three hundred SES leaders in all areas of operations. About 80 percent of the department's SES billets are in Career positions; the remaining 20 percent hold General noncareer positions. The Department of the Navy will continue to seek and develop their senior civilian leaders from within their workforce and reach out to industry, academia, or any source to fill leadership needs. The SES program provides the framework to achieve both stability and agility in leading the Department of the Navy's civilian workforce.

QUICKREFS

- Military personnel have both *billets* (current jobs) and *ranks* (qualified levels that grant pay, authority, and responsibility).
- All military personnel have *paygrades* and, though the pay at each level is the same for all services, different services use different names (ranks) to identify each paygrade.
- There are *enlisted* (E-1 through E-9) and *officer* (O-1 through O-10) paygrades in all the services. Warrant officers (W-1 through W-5) come between the officer and enlisted ranks.

- "Sailor" nowadays is a generic term that applies to *all* Navy uniformed personnel (officer and enlisted).
- In the Navy (and Coast Guard), but not the other services, enlisted personnel ranks are called "rates."
- In the Navy (and Coast Guard), but not the other services, enlisted personnel also have *ratings*, which are occupational specialties.
- There are different abbreviations and different ways of addressing military people, and these are all summarized in tables 1.2 through 1.6.
- When identifying or addressing military personnel, the letters and numbers *in front of* their names are what matter most.
- Officer ranks are not complicated, but you will need to familiarize yourself with them by reviewing the tables.
- Enlisted titles in the Navy are somewhat complicated because they frequently include the abbreviation for that person's rating (GM, QM, ABE, etc.) and there are too many of these to memorize. To properly address enlisted Sailors, these rating abbreviations *can be ignored* if you look at the "alphabet soup" preceding a Sailor's name to find the following clues, either standing alone or (more often) at the end of the letters:
  SA or SN = Seaman
  FA or FN = Fireman
  AA or AN = Airman
  CA or CN = Constructionman
  HA or HN = Hospitalman
  1, 2, or 3 = Petty Officer
  C = Chief
  CS = Senior Chief
  CM = Master Chief
- Part of the management team that makes up the leadership of the Navy is a cadre of civilians officially referred to as the Senior Executive Service (SES). While they wear no rank, these individuals are the equivalent of admirals and generals and should be accorded the same courtesies and protocols.

CHAPTER TWO

Navy Organization

The Navy's organization is unique, and understanding it can be challenging. In today's Navy, where "jointness" is a reality, it is not enough to merely understand how the Navy is organized; you must also have some idea of how it fits into the Department of Defense. This chapter will unravel some of the complexity and give you a better understanding of how the Navy is able to carry out its many assigned tasks.

COMPLICATIONS

Before trying to understand how the U.S. Navy is organized, you should be forewarned that it is complicated. One might think that setting up a navy organization would be a relatively simple thing; that ships would be organized into fleets to operate in certain waters of the world and that admirals would command those fleets; that a chain of command could be simply drawn from the commanding officers of ships to the commanders of fleets and ultimately to the senior-most admiral in charge of the whole navy. But this simplistic vision ignores a number of facts, and the reality is that there are actually a number of chains of command that must be understood if you are going to understand how the U.S. Navy is organized.

To begin with, the Navy consists of more than ships. There are also aircraft and submarines, SEALs, SeaBees, Marines, and so on that make up what we can collectively describe as the "operating forces" of the Navy.

Further, these operating forces cannot function independently. There must be a supply system to ensure that the operating forces have fuel, ammunition, and the like. Some means of repairing the ships, aircraft, and so on must be in place, and medical facilities must be available to care for battle casualties and the sick. These and other considerations mean that there must be facilities ashore to support those ships and aircraft.

Another complication stems from the realization that navies rarely operate alone, that modern warfare and readiness for war require all of the armed forces to operate together—or jointly—in various ways.

One more complicating factor comes from the fact that our nation is a democracy, and one of our governing principles is civilian control of the military.

These essential factors—the operating forces needing support from ashore, the need for joint operations among the services, and the necessity for civilian control of the armed forces—all combine to make for a more complicated organization than we might wish for, but understanding that organization can be helped by keeping the following in mind:

- In the military, the term "chain of command" is roughly synonymous with "organization." The former is a path of actual legal authority, whereas the latter is a little less formal, but for most purposes the two terms can be considered the same.
- There are two separate (but sometimes overlapping) chains of command within the Navy:
  — the operational chain of command, which is "task oriented," meaning that it is concerned with carrying out specific missions (combat operations, fleet exercises, humanitarian operations, etc.), and
  — the administrative chain of command, which takes care of matters like personnel manning, education and training, repairs, supply, and so on.
- The operating forces are permanently organized in an administrative chain of command, though they are frequently reassigned to different operational chains of command as needs arise.
- Even though the operational forces consist of more than just ships, they are often collectively referred to as simply "the Fleet."
- The Fleet is supported by a number of different commands and organizations known collectively as "the shore establishment."

- The Department of the Navy (DON)* is an integral part of the Department of Defense (DOD), which also includes the Army and the Air Force, and DON and DOD are intertwined to a significant degree.
- There is a civilian head of the Navy, known as the Secretary of the Navy (SECNAV), and a military head, the Chief of Naval Operations (CNO), and CNO is subordinate to SECNAV.
- The U.S. Marine Corps is a part of the DON but is in many ways a separate service as well, having its own senior military commander (known as the Commandant of the Marine Corps) but answering to the same civilian official (the SECNAV).
- There are also allied chains of command that sometimes must be considered. For example, because the United States is a key member of the North Atlantic Treaty Organization (NATO), a U.S. Navy admiral can be the NATO Supreme Allied Commander Europe and be responsible for forces belonging to member nations as well as those of the United States.
- Some individuals in this organizational structure may "wear two hats," an expression that means one person can actually have more than one job, and often those two (or more) jobs might be in different parts of the organizations described above. For example, the Commander of the Navy's 5th Fleet also currently holds the position of "Commander of U.S. Naval Forces, Central Command" (a joint command position within the DOD chain of command). The CNO "wears a number of hats" in that he or she is responsible for ensuring the readiness of the operating forces of the Navy but is also the head of the shore establishment, and though this admiral is the senior military officer in the Navy, he or she also works directly for the civilian SECNAV in many matters and serves as a member of the Joint Chiefs of Staff in matters that involve working with the other armed services.

Some final thoughts in dealing with the complicated nature of the Navy's organization. One is that you probably will not need to know every detail of

---

*While this is not terribly significant, to avoid any possible confusion you should know that according to *Navy Regulations,* there is a difference between the "Department of the Navy" (refers to the entire Navy organization—all operating forces including the Marine Corps) and the "Navy Department" (located at the seat of government in Washington D.C., it is comprised of the Office of the Secretary of the Navy, the Office of the Chief of Naval Operations, and Headquarters, Marine Corps). which are located in Washington, D.C.

that organization. The information provided below is there for an encompassing overview. Try to grasp the essentials—particularly those that apply directly to your job—but do not worry if you cannot remember exactly how it all fits together. Few people can.

Something else to keep in mind is that this organization frequently changes. Commands are renamed, offices shift responsibilities. Some confusion may be avoided if you do not always assume that what you remember is still the same. Also be wary of the Internet; though it is a wonderful information tool, it must be used with caution. Web sites are not always kept up-to-date, and older (out of date) items will show up when using a general search engine.

Keeping the above explanations in mind, let us venture into the labyrinth of Navy organization, beginning first with a description of the larger Department of Defense to put the Navy into proper perspective.

DEPARTMENT OF DEFENSE

As discussed above, the Department of the Navy (DON) is part of the Department of Defense (DOD), and some of the Navy's organization is directly intertwined with the DOD joint command structure. In its simplest breakdown, there are four principal components to DOD:

- the Secretary of Defense and his or her supporting staff
- the Joint Chiefs of Staff and their supporting staff
- the individual military departments (services): Army, Air Force, and Navy
- the Unified Combatant Commands

Most people who watch even a little news are aware that there is a Secretary of Defense (SECDEF) heading up DOD and that he or she is assisted by a senior military officer known as the Chairman of the Joint Chiefs of Staff (CJCS), who can come from any of the services and whose principal duties include advising the President, the National Security Council, and the SECDEF. Those people who are more informed may also know that the senior military officers of each service (the CNO; the Commandant of the Marine Corps, the Chief of Staff of the Army; and the Chief of Staff of the Air Force) serve collectively as the Joint Chiefs of Staff (JCS) under the Chairman. Note that though the Marine Corps is a service within the DON, the Commandant is a member of the JCS. (As previously noted: "complicated!")

The Coast Guard is another unique entity. It is, by law, the fifth military branch of the U.S. armed services, but it is assigned to the Department of Homeland Security, rather than DOD. And though the Coast Guard frequently operates in support of the Navy and DOD when it is called upon to perform national defense missions, the Commandant of the Coast Guard is not a formal member of the JCS. During wartime or national emergency, the President can have the Coast Guard assigned to the DON, but the last time this transfer occurred was just before and during World War II, and it is unlikely that it will ever happen again.

Both the Secretary of Defense and the Chairman of the Joint Chiefs of Staff have fairly large organizations working for them. Supporting SECDEF is a staff structure known collectively as the Office of the Secretary of Defense, which includes a Deputy Secretary of Defense, a number of undersecretaries, assistant secretaries, and other officials in charge of specific aspects of running DOD (see figure 2.1).

Likewise, the Chairman of the Joint Chiefs of Staff also has a support organization called "The Joint Staff," part of which is shown in figure 2.2. Note that specific areas of responsibility are assigned "J-codes," J-1 for Manpower and Personnel, J-2 for Intelligence, and so on. So if you hear someone say something like, "Smith over in J-4 needs to meet with us in the morning," you will know that the meeting will probably have something to do with logistics. The individual services mimic this system (with modifications) but use other letters. Intelligence on Army staffs is usually "G-2" and on Navy staffs, it is "N-2," for example.

For operational matters—such as contingency planning, responding to an international crisis, going to war, or participating in a major joint operational exercise—the chain of command is a bit different. This operational chain of command is sometimes referred to as the "U.S. National Defense Command Structure" and begins with the President of the United States in his or her constitutional role as Commander in Chief of the Armed Forces. It then goes through the SECDEF (with the CJCS* and the service chiefs serving as

---

*Some confusion sometimes arises as to the role of the CJCS in the operational chain of command. The Chairman (with the assistance of the JCS) technically serves only as a principal advisor to the SECDEF, but he or she often issues directives to the unified commanders, which sometimes creates the illusion that CJCS is in the chain of command between SECDEF and the unified commanders; however, these directives are always issued with the understanding that they originate with SECDEF, not from CJCS.

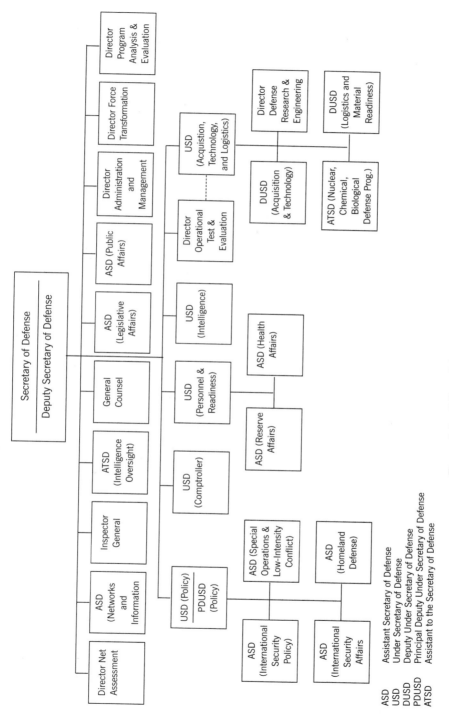

Figure 2.1 Office of the Secretary of Defense

ASD   Assistant Secretary of Defense
USD   Under Secretary of Defense
DUSD   Deputy Under Secretary of Defense
PDUSD   Principal Deputy Under Secretary of Defense
ATSD   Assistant to the Secretary of Defense

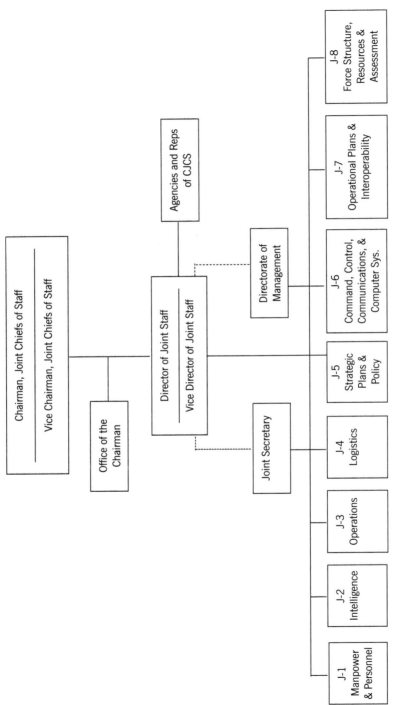

Figure 2.2 The Joint Staff

principal advisors) to those generals and admirals known as "unified combatant commanders" (sometimes referred to as just "unified commanders" and occasionally "combatant commanders") (see figure 2.3). Each is responsible for a specific geographic region of the world (sometimes referred to as an Area of Responsibility or AOR) or has a worldwide functional AOR. It is through the unified commanders that the DOD and Navy operational organizations come together.

Unified Combatant Commanders

There are currently nine unified commanders. Five are responsible for specific geographic regions of the world (sometimes referred to as an Area of Responsibility or AOR): Northern Command, Southern Command, Pacific Command, European Command, and Central Command.* Their AORs are shown in figure 2.4.

The other four have worldwide responsibilities that are functionally, rather than geographically, oriented: Joint Forces Command, Strategic Command, Transportation Command, and Special Forces Command.

All unified commanders answer directly to the Commander in Chief (the President of the United States) through the SECDEF.

U.S. Northern Command (USNORTHCOM or NORTHCOM)

One of the newer unified commands, NORTHCOM was created as a result of the attacks on U.S. soil on 11 September 2001. Its primary purposes are to coordinate the homeland defense missions that are carried out by DOD activities and to provide military assistance to civil authorities as needed.

Headquarters: Peterson Air Force Base, Colorado Springs, Colorado

Geographic responsibility: Continental United States, Alaska, Canada, Mexico, Puerto Rico, U.S. Virgin Islands, and surrounding waters out to five hundred nautical miles. Includes the Gulf of Mexico and some of the Caribbean, although most of the latter is under Southern Command (see below). Hawaii,

---

*To avoid any possible confusion, you should be aware that the formal titles of these commands include the prefix "United States," so "Northern Command (NORTHCOM)" is actually "United States Northern Command (USNORTHCOM)," but they are frequently referred to without the "US" prefix for brevity. We will use both in this discussion, but you should understand that they are one and the same.

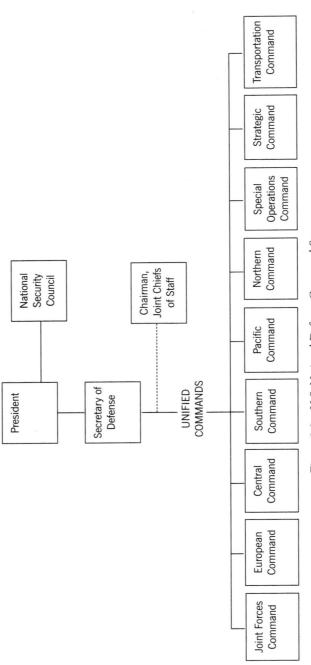

Figure 2.3 U.S. National Defense Command Structure

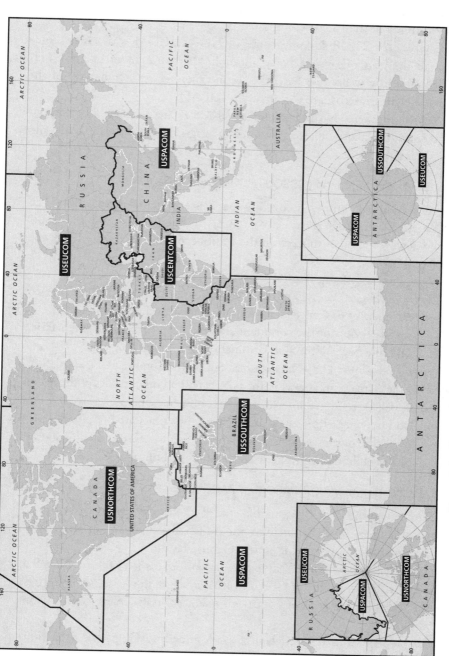

Figure 2.4 Geographic Areas of Responsibility for Unified Commands

Guam, and other U.S. territories and possessions in the Pacific are the responsibility of the Pacific Command.

Commander: The first commander of NORTHCOM was an Air Force general; the next one was a Navy admiral. The commander of NORTH-COM currently wears two hats, the second one being Commander of the North American Aerospace Command (NORAD).

Naval component: No permanent naval forces are assigned, but components from the Atlantic Fleet are frequently assigned for training and advisory functions as needed.

U.S. Southern Command (USSOUTHCOM or SOUTHCOM)

The other unified command covering the Western Hemisphere is SOUTH-COM. It encompasses thirty-two nations (nineteen in Central and South America with the other thirteen in the Caribbean) and covers nearly 15 million square miles.

Headquarters: Miami, Florida

Geographic responsibility: The land mass of Latin America south of Mexico, the waters adjacent to Central and South America, and the Caribbean.

Commander: Most of the commanders have been Army with one Marine general and one Navy admiral in command as of this writing.

Naval component: U.S. Naval Forces Southern Command (USNAVSO). Serves as the primary link between the U.S. Navy and other navies of the region for combined exercises, counter-drug operations, humanitarian missions, and so on. No permanent fleet assets are assigned, but units are periodically provided for specific purposes, such as the annual "UNITAS" multinational exercises aimed at building hemispheric cooperation and mutual defense.

U.S. Pacific Command (USPACOM or PACOM)

Encompassing the largest geographic area of all the regional unified combatant commanders, PACOM includes the Indian Ocean as well as the Pacific.

Headquarters: Camp H. M. Smith, Oahu, Hawaii

Geographic responsibility: Pacific and Indian oceans (less the Arabian Sea, which is the responsibility of CENTCOM), much of the Asian landmass (less Russia and some portions of Southwest Asia), Australia, New Zealand, and Antarctica.

Commander: Since 1947, when it was founded, PACOM has been commanded by a naval officer.

Naval component: The Pacific Fleet (which includes the 3rd Fleet and 7th Fleet).

U.S. Central Command (USCENTCOM or CENTCOM)

A direct descendant of what was once called the "rapid deployment force," which had been primarily a contingency force, CENTCOM has become one of the most active combatant commands, having responsibility for combat operations in Afghanistan and Iraq.

Headquarters: MacDill Air Force Base, Tampa, Florida

Geographic responsibility: Roughly the Middle East, Southwest Asia (less Israel, Lebanon, and Syria, which are the responsibility of EUCOM), and northwest Africa. Includes Iran, Iraq, Saudi Arabia and the other Gulf states, Egypt, Sudan, Eritrea, Ethiopia, Somalia, Kenya, Pakistan, Afghanistan, Uzbekistan, Turkmenistan, Tajikistan, Kazakhstan, and Kyrgyzstan, as well as the Persian Gulf and Arabian Sea.

Commander: So far, all CENTCOM commanders have been Army or Marine generals and one Navy admiral.

Naval component: U.S. Naval Forces Central Command (which includes the 5th Fleet).

U.S. European Command (USEUCOM or EUCOM)

Those areas not covered by the previous regional unified combatant commanders are covered by EUCOM. The commander also has allied responsibilities through NATO.

Headquarters: Stuttgart-Vaihingen, Germany

Geographic responsibility: Besides the continent of Europe, EUCOM includes Russia and several of the nations previously included in the now defunct Soviet Union (Ukraine, Belarus, Georgia, Moldova, Armenia, and Azerbaijan), part of the Middle East (Israel, Syria, and Lebanon), most of Africa (less Egypt, Sudan, Eritrea, Ethiopia, Somalia, and Kenya, which are the responsibility of CENTCOM), and the Mediterranean Sea.

Commander: EUCOM commanders have always been Army and Air Force generals until January 2003, when the command was given to a Marine general for the first time. This unified commander is also the commander of the NATO Supreme Allied Command Europe (SACEUR).

Naval component: Naval Force Europe, which includes the 6th Fleet.

U.S. Joint Forces Command (USJFCOM or JFCOM)

This specialized unified command is the "transformation laboratory" of the U.S. military, tasked with exploring promising alternatives in joint force usage, enhancing warfighting capabilities, and delivering trained and ready joint forces to warfighting commanders. A direct descendant of the old Atlantic Command (LANTCOM) and U.S. Atlantic Command (USACOM), this newest iteration has relinquished different portions of its former geographic responsibilities to NORTHCOM, CENTCOM, and EUCOM. However, the naval component commander (Commander U.S. Fleet Forces Command) maintains geographic responsibility for the entire Atlantic Ocean area, extending from the North Pole to the South Pole, the Caribbean Sea, and the waters around Central and South America extending in the Pacific to the Galapagos Islands.

Headquarters: Norfolk, Virginia

Commander: Traditionally commanded by Navy admirals (and one Marine general) when it was LANTCOM and USACOM, this commander also wore a second hat as the Supreme Allied Commander, Atlantic; the latter role has been replaced by a new command, now called Allied Command Transformation with a similar role in transforming NATO.

Naval component: U.S. Fleet Forces Command (which includes the 2nd Fleet).

U.S. Strategic Command (USSTRATCOM or STRATCOM)

As the command and control organization for all U.S. strategic forces (missile submarines, long-range bombers, ICBM systems, etc.), military space systems and operations, computer network systems, information operations, early warning systems, and strategic planning, STRATCOM is tasked with meeting both deterrent and decisive national security objectives. It includes components of the old Space Command (SPACECOM), the Air Force's Strategic Air Command (SAC), and the Navy's fleet ballistic missile submarine forces, among others.

Headquarters: Offutt Air Force Base, Nebraska

Geographic responsibility: Worldwide

Commander: Air Force, Navy, and Marine officers have all commanded STRATCOM.

Naval component: Submarines in Navy Task Forces 134 (Pacific) and 144 (Atlantic).

U.S. Transportation Command (USTRANSCOM or TRANSCOM)

The mission of this unified command is to provide air, land, and sea transportation for DOD, in both peace and war. TRANSCOM directly controls nearly ninety ships, more than a thousand aircraft, and more than two thousand railcars, and indirectly, though commercial partners, has access to many more transport assets. Component commands include the Air Force's Air Mobility Command (AMC), the Army's Surface Deployment and Distribution Command (SDDC), and the Navy's Military Sealift Command (MSC). Together, these commands meet the transport needs of DOD (including air refueling, aeromedical evacuation, land transport, ocean terminal management, commercial shipping interface, replenishment at sea, and a variety of other services that accomplish strategic and tactical airlift as well as move DOD personnel, their families, and their household goods worldwide).

Headquarters: Scott Air Force Base, Illinois

Geographic responsibility: Worldwide

Commander: To date, this command has always been headed by an Air Force general.

Naval component: MSC

U.S. Special Operations Command (USSOCOM or SOCOM)

Operations conducted by Special Forces (SEALS, "Green Berets," etc.) are coordinated by SOCOM through the subordinate commanders of the geographic unified commanders (PACOM, CENTCOM, etc.).

Headquarters: MacDill Air Force Base, Tampa, Florida

Geographic responsibility: Worldwide

Commander: SOCOM has been commanded by one Air Force general; all the others have been Army generals as of this writing.

Naval component: Naval Special Warfare Command

Specified Commands

You may also encounter the term "specified command." In the past there have been commands with status equal to unified combatant commands but were made up of elements from only one service; for example, for many years, there was a Military Airlift Command (MAC) that consisted of only Air Force personnel (in 1992 MAC was absorbed into the Transportation unified command, which has personnel assigned from the Army and Navy as well as the Air Force). These one-service commands were called "specified commands." Currently, there are no specified commands in DOD.

THE NAVY

As described above, the Navy is organized in two different (but related) ways at the same time: the *operational* chain of command and the *administrative* chain of command. Depending upon where you are assigned, you may be part of one or both of these organizations.

The operational chain of command controls forces (ships, aircraft, etc.) that are assigned to combat operations, operational readiness exercises, humanitarian relief missions, evacuations, and so on or are on station carrying out missions like sea control, deterrence, and so on.

The administrative chain of command is what keeps the Navy functioning on a day-to-day basis so that the ships, aircraft, and so on are able to carry out operational tasks when assigned. This is the chain of command that takes care of the less colorful but essential elements of preparedness, such as training, repair, supply, personnel assignment, intelligence support, communications facilities, weather prediction, medical treatment, and so on.

Both of these chains of command have the President of the United States at the top as Commander in Chief. Below the President is the Secretary of Defense, with the Joint Chiefs of Staff as his or her principal advisors. Below SECDEF, the chains are different. For operational matters, the SECDEF issues tasking orders directly to the unified commanders, and for administrative matters, SECDEF relies upon the Secretary of the Navy to keep the Navy (and Marine Corps) ready. The rest of these chains of command are explained below.

The Operational Chain of Command

In centuries past, naval warfare could be effectively waged more or less independently, but in the modern age the importance of joint warfare cannot be overemphasized. In the vast majority of modern operations, whether they are combat, humanitarian, readiness, deterrent, or specialized, several or all of the U.S. armed forces must cooperate, coordinate, and combine their forces and plans to maximize their effectiveness and ensure mission accomplishment.

The Navy's operational chain of command is headed by the appropriate unified combatant commanders described above. The chain becomes purely naval below the unified commander with "naval component commanders" exercising operational control over one or more of the "numbered fleet commanders."* These massive fleets are then subdivided into "task forces" for specific operational requirements, and task forces can then be broken down into "task groups" with subordinate "task units," and so on. See figure 2.5 for one example of this operational chain of command from one ship up through the unified combatant commander. Keep in mind that this shows only one path up the chain—there would also be other ships assigned to the various task elements, units, and so on, and there might be other task components (such as a Task Element 76.1.1.2 and Task Groups 76.2 and 76.3, etc.). Also keep in mind that the chain depicted in figure 2.5 exists for some specific operational task; as of this writing, Task Force 76 shown in the figure is responsible for carrying out amphibious operations in the Pacific operating area. This example chain might have been created to have USS *Fort McHenry* deliver Marines to one of several key locations in the Pacific as part of a larger operation.**

Component Commanders

Working directly for the unified commanders are the component commanders. These are commanders who control personnel, aircraft, and so on from their individual services that can be made available to the unified commander for a joint operation. For example, the joint CENTCOM commander has the following service component commands assigned to carry out those operations that fall within CENTCOM's AOR:

- U.S. Army Central Command (USARCENT)
- U.S. Central Command Air Forces (USCENTAF)
- U.S. Naval Forces Central Command (USNAVCENT)
- U.S. Marine Forces Central Command (USMARCENT)

Each unified commander has a similar array of component commanders.

Once a unified commander has determined what assets (troops, ships, aircraft, etc.) he or she will need to carry out a specific mission, that commander

---

*Component and fleet commanders are explained below.
**Task organization is explained in more detail below.

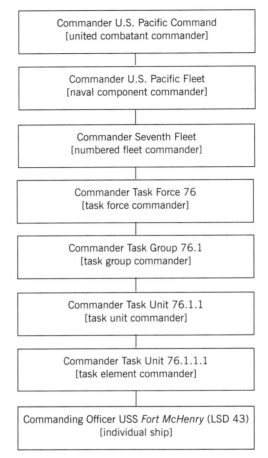

Figure 2.5   Naval Operational Chain of Command

will rely upon the component commanders to provide those forces and to coordinate their actions.

As already stated, this is the first level of command in the joint forces structure that is purely naval. There are a number of these naval component commands to meet the needs of the various regional unified commanders.

U.S. PACIFIC FLEET (USPACFLT OR PACFLT)

This naval component commander primarily serves the naval needs of the PACOM unified commander and the primary AOR is the same as that of

PACOM (Pacific Ocean and Indian Ocean), but PACFLT also provides assets to CENTCOM, SOUTHCOM, EUCOM, and STRATCOM when required. PACFLT is commanded by a four-star admiral, headquartered at Pearl Harbor, Hawaii, and exercises control of both the 3rd Fleet and 7th Fleet.

U.S. NAVAL FORCES EUROPE (USNAVEUR OR NAVEUR)

From headquarters in Naples, Italy, Commander of U.S. Naval Forces in Europe (COMUSNAVEUR) plans, conducts, and supports naval operations in the European AOR during peacetime, crisis, or war, answering directly to the EUCOM unified commander. He or she is supported by the Commander of the 6th Fleet and by the Commander, Navy Region Europe (both are also headquartered in Naples, Italy). COMUSNAVEUR also wears a second hat as NATO's Commander, Joint Force Command Naples.

U.S. NAVAL FORCES CENTRAL COMMAND (USNAVCENT OR NAVCENT)

Serving as naval component commander for the U.S. Central Command, NAVCENT is responsible for naval activities in the Arabian Sea, Persian Gulf, Red Sea, and part of the Indian Ocean. The vice admiral in command of this component also wears a second hat as Commander of the 5th Fleet.

U.S. FLEET FORCES COMMAND (USFLTFORCOM OR FLTFORCOM OR FFC)

Headquartered in Norfolk, Virginia, the Commander of Fleet Forces Command answers to both the Commander of Joint Forces Command (COMJFCOM) and the Commander of Northern Command (COMNORTHCOM) in addition to providing some naval support to the European Command (EUCOM) and the Central Command (CENTCOM). The 2nd Fleet is under the direct control of this component commander (who used to be known as the Commander of the Atlantic Fleet or COMLANTFLT). To avoid redundancies and to ensure consistency, he or she also coordinates personnel, training, maintenance, and administration matters between the Atlantic and Pacific fleets, ensuring that both fleets' procedures are the same.

Numbered Fleet Commanders

Commanding the ships, submarines, and aircraft that operate in direct support of the naval component commanders are vice admirals in charge of the numbered fleets. Like the component commanders, these commanders have support staffs and facilities ashore, but the numbered fleet commanders also have a flagship from which to conduct operations at sea when a mission calls for it. Individual ships, submarines, and aircraft squadrons are assigned to different fleets at different times during their operational schedules. For example, a destroyer assigned to 2nd Fleet while operating out of its home port of Norfolk might later deploy to the 6th Fleet in the Mediterranean for several months. There are currently six numbered fleets in the U.S. Navy.

Fleet	Primary Operational Area
2nd	Atlantic Ocean
3rd	Eastern Pacific Ocean
4th	Caribbean and South American Waters
5th	Middle Eastern waters
6th	Mediterranean Sea
7th	Western Pacific Ocean/Indian Ocean

The apparent gap (no 1st fleet) is an example of historical evolution rather than an oversight. Numbered fleets come and go according to the current needs. For example, in World War II there were 8th, 10th, and 12th fleets to meet specific needs of that global conflict, but they have since been deactivated.

2ND FLEET

With shore headquarters at Norfolk, Virginia, 2nd Fleet operates primarily in the Western Atlantic, but it also has Atlantic-wide responsibilities as a NATO strike force when needed.

3RD FLEET

With shore headquarters in San Diego, California, 3rd Fleet operates primarily in the Eastern Pacific and supplies units on a rotational basis to 5th Fleet in the Western Pacific Ocean and Indian Ocean and to 5th Fleet in the Middle East.

4TH FLEET

Headquartered in Mayport, Florida, 4th Fleet is responsible for the Caribbean and the waters around Central and South America.

5TH FLEET

With shore headquarters in Bahrain on the Persian Gulf, 5th Fleet has operational control of those units operating in the CENTCOM AOR.

6TH FLEET

Operating in the Mediterranean Sea with shore headquarters in Gaeta, Italy, 6th Fleet has both U.S. and NATO responsibilities (the latter as components of the NATO Strike and Support Forces, Southern Europe). Units assigned to 6th Fleet come primarily from 2nd Fleet on a rotational basis.

7TH FLEET

With shore headquarters in Yokosuka, Japan, 7th Fleet is responsible for the Western Pacific Ocean and Indian Ocean. The majority of units come from 3rd Fleet on a rotational basis, but there are some permanently assigned assets in the 7th Fleet that are homeported in Japan.

Task Organization

An entire fleet is too large to be used for most specific operations, but a particular task may require more than one ship. To better organize ships into useful groups, the Navy developed an organizational system that has been in use since World War II.

Using this system, a fleet can be divided into task forces and they can be further subdivided into task groups. If these task groups still need to be further divided, task units can be created and they can be further subdivided into task elements.

A numbering system is used to make it clear what each of these divisions is. The 7th Fleet, for example, might be divided into two task forces and they would be numbered TF 71 and TF 72. If Task Force 72 needed to be divided into three separate divisions, they would be task groups and would be numbered TG 72.1, TG 72.2, and TG 72.3. If TG 72.3 needed to be subdivided, it could be broken into task units numbered TU 72.3.1 and TU 72.3.2 (this "decimal" system might not sit well with your high-school math teacher, but it works for the Navy). Further divisions of TU 72.3.1 would be elements and would be numbered TE 72.3.1.1 and TE 72.3.1.2. This system can be used to create virtually any number of task forces, groups, units, and elements, limited only by the number of ships available.

The Administrative Chain of Command

As mentioned above, there is a chain of command within the Navy that is parallel to the operational one; this one involves many (but not all) of the same people who "wear more than one hat" in order to carry out functions within each chain. The administrative chain is concerned with readiness more than execution, focusing on such vital matters as manning, training, and supply, so that the operating forces are prepared to carry out those missions assigned by the operational chain of command.

Secretary of the Navy

SECNAV has an Undersecretary of the Navy as his or her direct assistant and, as can be seen in figure 2.6, there are several Assistant Secretaries of the Navy to handle specific administrative areas of the Navy.

In addition to these civilian assistants, SECNAV also has a number of military assistants, such as the Navy's Judge Advocate General, the Naval Inspector General, and the Chief of Information. Other military officers answer directly to SECNAV's civilian assistants, such as the Chief of Naval Research who reports to the Assistant Secretary of the Navy for Research, Development, and Acquisition, and the Auditor General who reports to the Undersecretary of the Navy.

Chief of Naval Operations

Reporting directly to the SECNAV is the Chief of Naval Operations.* The CNO is the senior military officer in the Navy and as such, he or she is a member of the JCS (discussed above) and is also the principal advisor to the President and to the SECDEF for naval matters. The CNO is always a four-star admiral and is responsible to the SECNAV for the manning, training, maintaining, and equipping of the Navy, as well as its operating efficiency. Despite the name, he or she is not in the operational chain of command.

OFFICE OF THE CHIEF OF NAVAL OPERATIONS
Besides a Vice Chief of Naval Operations (VCNO), the CNO has a number of admirals working for him or her who oversee specific functions within the Navy as indicated in figure 2.7. Most of them identified by the title "Deputy Chief of Naval Operations" (DCNO), the exception being the Director of Naval Intelligence.

Collectively, the CNO's staff is commonly referred to as "the Navy Staff" or "OPNAV" (derived from "operations of the Navy" but better thought of as simply the "Office of the Chief of Naval Operations"). OPNAV also assists SECNAV, the Undersecretary, and the Assistant Secretaries of the Navy.

---

*The Commandant of the Marine Corps also reports directly to SECNAV.

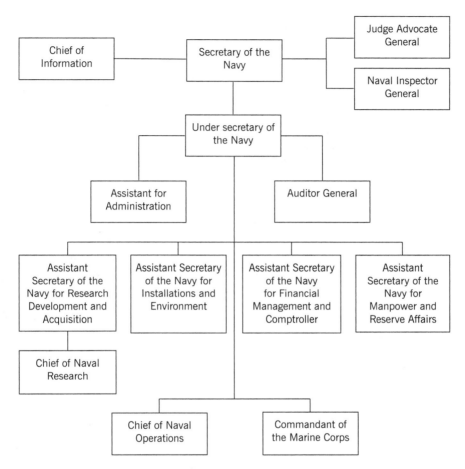

Figure 2.6 The Secretary of the Navy and his or her assistants.

You might have noticed that the CNO and his or her assistants have "N" codes that are similar to the "J" codes we encountered on the Joint Chiefs staff (see figure 2.2); for example, on the Joint Staff, "J-1" identifies the head of Manpower and Personnel and the same is true for the Navy's "N-1." "J-2" and "N-2" are responsible for Intelligence and so on. There are a few differences, however. Though there are separate J-3 (Operations) and J-5 (Strategic Plans and Policy) codes on the Joint Staff, the OPNAV Staff currently combines these functions into "N3/N5" (Plans, Policy, and Operations). Also, the Joint Staff has a J-7 (Operational Plans and Joint Force Development) and a

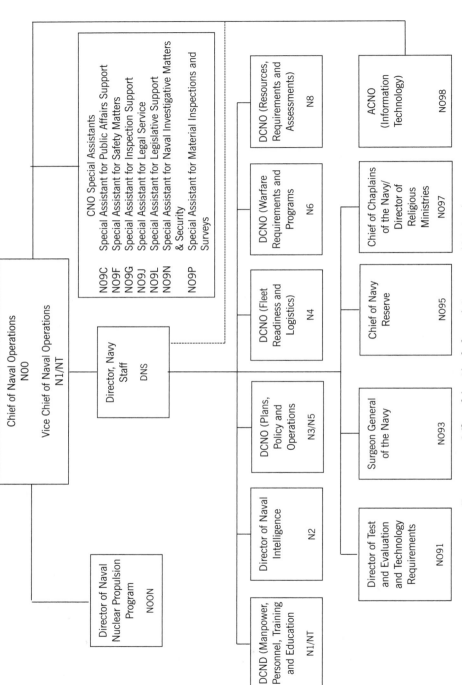

Figure 2.7 Office of the Chief of Naval Operations (OPNAV)

J-8 (Force Structure, Resources, and Assessment), but the OPNAV Staff does not have an N7, and the N8 is defined as "DCNO for Integration of Capabilities and Resources."

These admirals also have large staffs and subordinate commanders working for them, with many of these officers identified by subordinate N-codes as well. For example, the vice admiral who is assigned as DCNO for Integration of Capabilities and Resources has a number of rear admirals working for him or her as N85 (Expeditionary Warfare), N86 (Surface Warfare), N87 (Submarine Warfare), N88 (Air Warfare), and so on.

There are numerous other assistants to the CNO on the OPNAV Staff, such as the Special Assistant for Legislative Support (N09L), the Director of Naval Education and Training (N00T), the Surgeon General of the Navy (N093) who oversees all medical activities within the DON, the Chief of Chaplains (N097), and the Director of Navy Reserve (N095).

The Director, Navy Staff directs OPNAV Staff Principal Officials in support of CNO executive decision making, delivers management support to the OPNAV Staff, and serves as sponsor for thirty of the Navy's most important naval commands.

SHORE ESTABLISHMENT

In addition to the OPNAV Staff, there are a number of shore commands directly under the CNO that support the Fleet, including the Office of Naval Intelligence, the Naval Security Group Command, the Naval Safety Center, the Naval Meteorology and Oceanography Command, the Naval Strike and Air Warfare Center, the Naval Legal Service Command, the Bureau of Naval Personnel, the Naval Education and Training Command, and the Bureau of Medicine and Surgery. A number of these commands are dual-hatted: for example, the head of the Bureau of Naval Personnel (known officially as the "Chief of Naval Personnel") is also the Deputy CNO for Manpower and Personnel (N1), and the Commander Naval Education and Training Command serves as a member of the OPNAV Staff, advising the CNO as the Director of Naval Education and Training (N00T).

SYSTEMS COMMANDS

There are also five systems commands that oversee many of the technical requirements of the Navy and report to the CNO and SECNAV.

Naval Sea Systems Command (NAVSEA) is the largest and serves as the central activity for the building of ships, their maintenance and repair, and the procurement of those systems and equipments necessary to keep them operational. Among its many functions and responsibilities, NAVSEA also oversees explosive ordnance safety as well as salvage and diving operations within the Navy.

Naval Air Systems Command (NAVAIR) researches, acquires, develops, and supports technical systems and components for the aviation requirements of the Navy, Marine Corps, and Coast Guard.

Space and Naval Warfare Systems Command (SPAWAR) is responsible for the Navy's command, control, communications, computer, intelligence, and surveillance systems. These systems are used in combat operations, weather and oceanographic forecasting, navigation, and space operations.

Naval Supply Systems Command (NAVSUP) provides logistic support to the Navy, ensuring that adequate supplies of ammunition, fuel, food, repair parts, and so on are acquired and distributed worldwide to naval forces.

Naval Facilities Engineering Command (NAVFAC) is responsible for the planning, design, and construction of public works, family and bachelor housing, and public utilities for the Navy around the world. NAVFAC manages the Navy's real estate and oversees environmental projects while keeping its bases running efficiently.

TYPE COMMANDS

For administrative purposes (personnel manning, training, scheduled repairs, etc.), the Pacific and Atlantic fleets have ships and aircraft classified and organized into commands related directly to their type. These groupings are called, appropriately enough, "type commands" and there are six as follows:

- *Naval Surface Force, U.S. Atlantic Fleet (SURFLANT)* administers to the needs and ensures the readiness of all surface ships (cruisers and destroyers, as well as amphibious, service, and mine-warfare ships) assigned to the Commander of the U.S. Fleet Forces Command. This type commander is known as COMNAVSURFLANT and his or her headquarters are in Norfolk, Virginia.
- *Naval Surface Force, U.S. Pacific Fleet (SURFPAC)* administers to the needs and ensures the readiness of all surface ships assigned to the Commander of the U.S. Pacific Fleet. This type commander is known as COMNAV-SURFPAC and his or her headquarters are in San Diego, California.

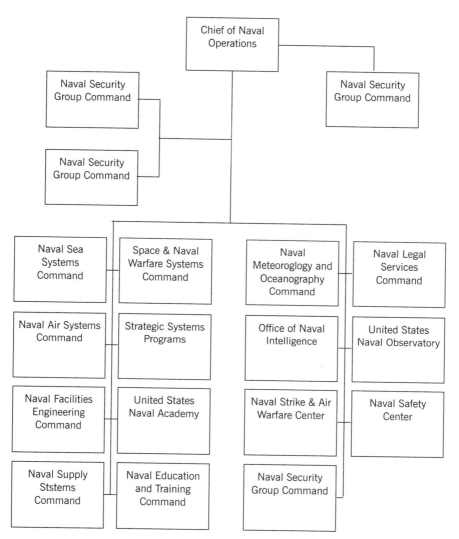

Figure 2.8   Major Components of the Navy Shore Establishment

- *Naval Submarine Force, U.S. Atlantic Fleet (SUBLANT)* administers to the needs and ensures the readiness of all submarines assigned to the U.S. Atlantic Fleet. This type commander is known as COMNAVSUB-LANT and his or her headquarters are in Norfolk, Virginia.

- *Naval Submarine Force, U.S. Pacific Fleet (SUBPAC)* administers to the needs and ensures the readiness of all submarines assigned to the Commander of the U.S. Pacific Fleet. This type commander is known as COMNAVSUBPAC and his or her headquarters are in Pearl Harbor, Hawaii.
- *Naval Air Force, U.S. Atlantic Fleet (AIRLANT)* administers to the needs and ensures the readiness of all aircraft assigned to the U.S. Atlantic Fleet. This type commander is known as COMNAVAIRLANT and his or her headquarters are in Norfolk, Virginia.
- *Naval Air Force, U.S. Pacific Fleet (AIRPAC)* administers to the needs and ensures the readiness of all aircraft assigned to the Commander of the U.S. Pacific Fleet. This type commander is known as COMNAVAIRPAC and his or her headquarters are in North Island, California.

Even though there are separate type commanders on the different coasts, for coordination purposes, one is senior to the other and ensures compatibility of resources and procedures. For example, COMNAVSURFLANT is a three-star admiral and COMNAVSURFPAC is a two-star; the latter in charge of surface activities in the Pacific, yet deferring to her or his counterpart on the Atlantic to ensure that the two are not too different. This was not always the case, so that Sailors moving from one coast to the other often encountered very different rules and procedures.

OTHER COMPONENTS TO THE ADMINISTRATIVE CHAIN
OF COMMAND

Below the type commanders are group commanders, and below them are ship squadron commanders or air wing commanders. For example, below COM-NAVSURFLANT is Commander Naval Surface Group 2 with Destroyer Squadrons (DESRON) 6 and 14. Do not look for any regular pattern in these numbering systems—except that components on the East Coast are generally even-numbered and components on the West Coast are odd-numbered—that is, just because Surface Group 2 has DESRONs 6 and 14 does not mean it will also have those numbers in between.

THE CHAINS OF COMMAND IN USE

To better understand these chains of command, let us look at an imaginary but typical situation from the deck plates up. Suppose you are assigned to a destroyer, USS *Neversink*, which is homeported in Norfolk, Virginia. Administratively, your ship's immediate superior in command is the Commander of Destroyer Squadron 8 (COMDESRON 8), along with five other destroyers. DESRONs 8 and 14 are assigned to Naval Surface Group 4, which is assigned to the Commander of Naval Surface Forces Atlantic (COMNAVSUR-FLANT). As indicated above, COMNAVSURFLANT answers to the Commander of Fleet Forces Command (CFFC).

While you are operating in and out of Norfolk—doing training exercises and getting important repairs completed—you would be operationally assigned to the Commander of the U.S. 2nd Fleet (COMSECONDFLT) as well. But when your ship deploys to the Mediterranean, you are no longer a member of the 2nd Fleet, but are transferred (operationally) to the Commander of the U.S. 6th Fleet (COMSIXTHFLT), who is the naval operational commander in the Mediterranean. He or she answers to COMUSNAVEUR (Commander of U.S. Naval Forces in Europe), who would be an admiral in charge of all naval forces in and around the European continent. That admiral would also have a boss—who might be another Navy admiral or could be a general from one of the other armed forces and is a unified combatant commander—whose title is COMUSEUCOM (Commander U.S. European Command), in charge of all U.S. armed forces in Europe (this includes Army and Air Force as well as Navy and Marine Corps). This commander would in turn answer to the SECDEF, who is supported and advised by the JCS, and his or her superior is the President of the United States (Commander in Chief of the Armed Forces).

Keep in mind that your ship's deployment to the Mediterranean, with all the changes that means to your operational chain of command, does not change your administrative chain of command. For personnel replacements, scheduling of repair overhauls, and the like, *Neversink* would still work through DESRON 8, Naval Surface Group 2, COMNAVSURFLANT, and so on.

Suppose that while your ship is in the Mediterranean, trouble flares up in a small coastal town and the lives of a number of American citizens are in jeopardy. The President orders the Secretary of Defense to ensure that no harm comes to those Americans. With the advice of the Joint Chiefs, SECDEF decides that evacuating the Americans is the only way to ensure their safety. Carrying out SECDEF's order, the Chairman of the Joint Chiefs sends a message to the appropriate unified commander, COMUSEUCOM, who in turn decides to use some of the Marines embarked in an Expeditionary Strike

Group operating in the vicinity to carry out the rescue. He tells COMUS-NAVEUR to make it happen, and she devises a plan to insert a contingent of Marines onto a beach just north of the town before dawn so that they can rescue the Americans, who are gathered at a walled school on the northern edge of the town.

A task organization is set up for the operation by the 6th Fleet Commander as follows. The commander of Task Force 62 (TF 62), the combat-ready ground force composed of a Marine Expeditionary Unit (MEU) of approximately eighteen hundred Marines, orders one of his amphibious transport docks (LPDs) to immediately refuel and then move into the area to insert two companies of Marines ashore as planned. All the ships, aircraft, and Marines to be used in the evacuation operation are assigned to a task group designated TG 62.7. Your ship, *Neversink*, and two other destroyers are assigned to a supporting task unit (TU 62.7.1) with the mission of escorting the LPD, and because your ship's commanding officer is senior to the other destroyer captains, she is designated as the commander of the task unit (CTU 62.7.1). A dock landing ship (LSD) and one destroyer are assigned by CTG 62.7 as a reinforcement force if needed, and they are assigned as components of TU 62.7.2 with orders to stand off just outside territorial waters until (if) they are needed.

The oiler needed to refuel the LPD will come from the Mediterranean Logistics Force (Task Force 63), but for this operation, she and an escort will be designated as a supporting task element of the evacuation operation with the designation of TE 62.7.1.1 until the refueling mission is complete. Once the evacuation is accomplished and the mission is complete in all respects, TG 62.7 will be disbanded and the participating units will return to their routine schedules, which for *Neversink* includes a port visit to Cannes on the French Riviera.

A similar situation would exist if you had been assigned to USS *Sailwell*, homeported in San Diego. Your ship would be administratively assigned to COMNAVSURFPAC and operationally to COMTHIRDFLT while operating in and out of San Diego. Once you deployed to the Western Pacific, however, you would change your operational chain of command to COMSEVENTH-FLT (who is responsible for the Western Pacific) and the chain would extend upward through COMPACFLT to COMUSPAC to SECDEF to the President. During your WESTPAC (Western Pacific) deployment, your ship could be assigned to different task organizations (like TU 77.2.6 or TE 72.1.1.1) as needs arise.

COMPLICATED BUT FUNCTIONAL

As you can see from the above, the Navy's organization is indeed complicated, but considering all that must be done to keep the world's most powerful Navy ready and able to carry out a very wide variety of missions, that organization is capable of meeting the many needs of the Fleet as it serves the much larger defense establishment that guards the nation's vital interests and keeps American safe.

N-codes and task unit designations with multiple decimal points may seem intimidating at first, but those who operate within them are soon comfortable with it all. Once you are assigned to a specific command somewhere in that vast organization, you will quickly learn those parts of the chains of command that are important to you, and you will soon be using the Navy's alphabet soup of terminology with the best of them.

QUICKREFS

- There is more than one chain of command in the Navy.
  - —The *operational* chain of command is used to carry out specific missions (like an airstrike or a humanitarian relief effort).
  - —The *administrative* chain of command takes care of support functions (like personnel manning, repairs, etc.).
  - —U.S. forces are often part of *allied* chains of command as well.
- The Navy's operational forces (known as "the Fleet") consist of ships, aircraft, submarines, SEALs, and so on.
- The *Department of the Navy*
  - —is headed by a civilian Secretary of the Navy, who is assisted by a military head known as the Chief of Naval Operations, and
  - —is subordinate to the Department of Defense (which is headed by a civilian Secretary of Defense, who is assisted by a military Chairman of the Joint Chiefs of Staff).
- The *Marine Corps* is part of the Department of the Navy but is in many ways a separate service as well.
- The *Coast Guard* is the fifth branch of the U.S. armed services and often works closely with elements of the Department of Defense, but it is currently organized as part of the Department of Homeland Security.

- An individual may have more than one job (called "wearing two hats").
- The Navy's *operational chain of command* is structured as follows (from top to bottom):
  —President of the United States (constitutionally the "Commander in Chief of the Armed Forces"),
  —Secretary of Defense,
  —one of the *unified combatant commanders* (such as PACOM) depending upon the locale and the mission,
  —appropriate *naval component commander* (such as PACFLT) where it becomes purely naval (as opposed to joint),
  —*numbered fleet commander* (such as COMSEVENTHFLT), and
  —down through a *task organization* that can consist of all or some of the following (in descending order):
    task force
    task group
    task unit
    task element
    individual unit (such as a ship, aircraft squadron, SEAL team, etc.).
- The *Joint Chiefs of Staff* (including the Chairman) serve as principal advisors to the President and Secretary of Defense but are not actually in the operational chain of command.
- There are nine *unified commands*:
  —Central Command (CENTCOM)
  —European Command (EUCOM)
  —Joint Forces Command (JFCOM)
  —Northern Command (NORTHCOM)
  —Pacific Command (PACOM)
  —Southern Command (SOUTHCOM)
  —Special Operations Command (SOCOM)
  —Strategic Command (STRATCOM)
  —Transportation Command (TRANSCOM)
- The *naval component commands* are:
  —PACFLT (Pacific Fleet)
  —NAVEUR (Naval Forces Europe)
  —NAVCENT (Naval Forces Central Command)
  —FFC (Fleet Forces Command)

- There are currently five *numbered fleet commands* in the Navy operationally supporting the naval component commanders:

Fleet	Primary Operational Area
2nd	Atlantic Ocean
3rd	Eastern Pacific Ocean
5th	Middle Eastern waters
6th	Mediterranean Sea
7th	Western Pacific Ocean/Indian Ocean

- The *administrative chain of command* keeps the Navy functioning on a day-to-day basis by taking care of the essential elements of preparedness like training, repair, supply, personnel assignment, medical treatment, and so on. Unlike the operational chain, it is purely naval, rather than joint. The administrative chain of command is structured as follows:
  —The SECNAV, who is a civilian and has a second in command known as the Undersecretary, as well as several Assistant Secretaries who handle specific areas, such as Manpower and Reserve Affairs.
  —Subordinate to SECNAV is the *Chief of Naval Operations* (CNO), who is a naval officer and has a second in command known as the Vice Chief of Naval Operations, as well as a number of deputies with specific areas of responsibility such as Naval Intelligence. The CNO and his staff are organized as the Office of the Chief of Naval Operations, better known as "OPNAV."
  —Below SECNAV and OPNAV is the *shore establishment*, consisting of various activities that support the operating fleet by handling such things as Training, Medicine, and Intelligence.
  —Five *systems commands* oversee many of the technical requirements of the Navy and report to the CNO:
    Naval Sea Systems Command (NAVSEA)
    Naval Air Systems Command (NAVAIR)
    Space and Naval Warfare Systems Command (SPAWAR)
    Naval Supply Systems Command (NAVSUP)
    Naval Facilities Engineering Command (NAVFAC)
  —For administrative purposes, ships and aircraft are organized into six *type commands* that reflect their commonality and their geographic location:
    Naval Surface Force, U.S. Atlantic Fleet (SURFLANT)
    Naval Surface Force, U.S. Pacific Fleet (SURFPAC)

Naval Submarine Force, U.S. Atlantic Fleet (SUBLANT)

Naval Submarine Force, U.S. Pacific Fleet (SUBPAC)

Naval Air Force, U.S. Atlantic Fleet (AIRLANT)

Naval Air Force, U.S. Pacific Fleet (AIRPAC)

—Within the type commands, ships and aircraft are further organized into manageable *group commands* with component *ship squadrons* and *air wings*, consisting of individual ships and aircraft squadrons.

CHAPTER THREE

Reading Uniforms

Uniforms—defined as the clothes worn by military personnel and the various accoutrements that go on them—serve a variety of purposes: identification (as someone in the armed services), standardization (as belonging to a specific service), differentiation (as a means of establishing where one fits in the hierarchy of the service), and recognition (of individual achievements). In civilian life, we often make reasonable assumptions about people based upon their attire—what they are wearing and where they are wearing it. Though this is not a foolproof system, noting that a person is wearing a three-piece suit or sweat pants or cowboy boots can often lead us to some reasonable conclusions about that individual. The same is true with military uniforms, even though the reasons are a bit different, and being able to "read" military uniforms can be most helpful when you are surrounded by them.

Once you become familiar with some of the characteristics and the variations in uniforms, you will be able to tell a great deal about individuals simply by observing what they are wearing. This makes for smoother professional relations and can serve as a good vehicle to social conversation: "I see you are an aviator; what kind of aircraft do you fly?"

One difference you will be able to see is whether a person is an officer or enlisted. A closer look may tell you such things as what the person's rank or rate is, how many years he or she has served in the Navy, what her or his occupational specialty is, and whether or not he or she has had any special assignments in the Navy or has achieved any special qualifications. If you are really good at reading uniforms, you will be able to tell some of the places in which a person has served and whether or not he or she has received special recognition while serving in one or more billets.

After you have absorbed the material in this chapter, you will be able to meet a woman in uniform and, without a spoken word between you, tell that she is a second class petty officer with at least four years of service whose occupational specialty is photography, that she has been on sea duty, and that she has been to the Middle East at least once. Even the great detective Sherlock Holmes might be impressed by that kind of deduction.

THE CLOTHES

There are a great many variations of uniform worn by all service members, depending upon where they are, what they are doing, and what time of year it is. There are uniforms designed for heavy-duty working aboard ship, others more appropriate for working ashore in an office, still others for ceremonial occasions, some for very formal evening wear, and nearly all have variations depending upon seasonal conditions. Covering all the variations is beyond the scope of this book, but we will provide some of the basics so that you will be a lot less confused the next time you encounter people in uniforms.

Different Services

Without going into all the variations, the following hints will take some of the mystery out of deciphering the variety of colors you will encounter when people from different services are nearby. Sometimes the differences are easy to discern but at other times it is more difficult.

Telling the Services Apart

This should be a lot easier than it is, but there are many variations and combinations that sometimes make discernment difficult. A few general characteristics can narrow the possibilities and help you identify which service you are encountering.*

One uniform worn by the Navy is known officially as "Service Dress Blue" but is affectionately called the "Crackerjack" uniform (white Sailor hat, top shirt with a flap on the back, neckerchief, and bell-bottom trousers) (Photo 3.1). It is very distinctive and not likely to cause much confusion. That's the good news; the bad news is that it is worn only by enlisted personnel in the bottom six paygrades and not as often as some of the other Navy uniforms.

---

*The military uses some odd terminology (for example, the equivalent of a suit or sport jacket is called a "blouse") but I will use more familiar terminology here to avoid confusion.

Photo 3.1 Navy Service Dress Blue ("Crackerjack")

When dressed formally, Marines are very distinctive. The "choker" (high collar with no necktie) collars on their blue uniforms are unique. Marine trousers with those choker blues will usually be lighter blue (many, but not all, with red stripes down the outer pant leg); an exception is that in summer, officers may wear white trousers with this uniform. Marine officers also occasionally wear choker collar uniforms that are all white, a color that is also worn by Coast Guard and Navy officers and chief petty officers (Photos 3.2 and 3.3). No other services have choker collars (except at service academies).

If a person is wearing white, black, or khaki from head to toe he or she is probably Navy (Photo 3.4).*

New Navy uniforms for E-6 and below incorporate both khaki and black (khaki shirts with black trousers or skirts).

Air Force and Coast Guard personnel both wear blue, but they are noticeably different shades. The Air Force blue has more gray in it and the Coast

---

*There are a few other all-white uniforms in the armed services (such as Marines in "choker" whites and Air Force medical personnel) but these are infrequently encountered. You are more likely to encounter Navy personnel dressed in all white.

Photo 3.2   Marines in "Choker" Uniform

Photo 3.3   Naval Officers Wearing "Choker" Whites

Photo 3.4 Navy Chief Petty Officer in Khaki

Guard blue is more of a royal blue. One dead giveaway is that Air Force suit jackets have silver buttons and Coast Guard suit jackets have gold.

The Army also has a dress uniform that is dark blue, but (a) they rarely wear it (mostly for very formal occasions), and (b) you can tell them from the Air Force and Coast Guard because the Army's version has trousers or skirts that are a lighter shade of blue than the jacket, and you can tell it from the Marines because the Army does not have a "choker" collar and has a necktie instead.

Light green shirts (whether long or short sleeve and with or without a suit jacket) are a dead giveaway that the person is Army.

Light blue shirts usually mean Air Force unless there are Coast Guard personnel around—they also wear a light blue shirt.

Naval officers and chief petty officers wear white shirts with some uniforms, with or without suit jackets or sweaters.

Both Marines and Army wear green suit jackets at times with matching trousers or skirts, but they are very different shades (Marines wear a much darker green) and the Army's jacket buttons are gold, whereas the Marines' jacket buttons appear black.

If a servicemember is wearing a multitude of colors all at the same time (khaki shirt, blue trousers, and a white cap), he or she is a Marine.

A khaki necktie means Marine; blue neckties (in slightly different shades) are worn by the Air Force and Coast Guard; black neckties are worn by both Army and Navy.

A black beret on a man is probably Army, but women in the Navy also wear them. There are also black and maroon berets in the Air Force worn by security and pararescue personnel, respectively. Green berets are definitely Army, and there are a few other beret variations that may confuse you.

Bottom line: If in doubt, *ask individuals what service they are in*—they are used to it and much prefer that you ask, rather than make assumptions such as "commercial airline pilot" or "doorman"!

Camouflage Uniforms

One example of uniforms that are difficult to identify are camouflaged fatigues. There are variations of these depending upon the location and purpose (for example, jungle or woodland camouflage is predominantly green, whereas desert camouflage is predominantly brown or tan, and urban camouflage is made up of blacks and grays). Some Sailors may wear one of those variations (a hospital corpsman attached to a Marine company in Iraq may wear desert camouflage, a Navy diver serving in Colombia may be in jungle camouflage, etc.), or they may wear a newer version that is predominantly blue and gray for shipboard wear.

All the services wear these various camouflage uniforms at different times. The bad news is that they are often hard to tell apart from any distance; the good news is that if you get up close, they usually have their service name embroidered or stenciled over one of the pockets.

Officers

Though there is no quick, absolute way to tell officers by their uniforms, there are clues.*

Rank Devices

The most reliable way to tell if an individual is an officer is by the actual device worn somewhere on the uniform that tells you his or her specific rank. See *Navy, Army, Air Force, Marine Corps,* and *Coast Guard Officer Ranks* (p. 73).

Keep in mind that rank indications can appear in different places on the uniform, depending upon which uniform is being worn. For example, officers of all the services wear rank devices on their collars when wearing camouflage, but when wearing their standard dress winter uniforms Navy and Coast Guard officers wear their ranks on the bottom of the sleeve (near the cuff), whereas the other services wear their ranks on the shoulders of equivalent uniforms.

Among the various rank indicators worn by officers, collar devices are the ones common to all (though not worn on every uniform). On page 73 these are the bars, oak leaves, eagles, and stars (some in gold and silver variants). When worn, these are the same for all officers of all services, although they do actually appear slightly different in some instances (for example, Army collar devices are noticeably larger than those worn by the other services). Whether the lowest ranking officers are called "ensigns" (in the Navy and Coast Guard) or "second lieutenants" (in the other services), they will wear gold bars on their collars (if the particular uniform they are wearing requires a collar device). If you see a person wearing a gold bar anywhere on his or her uniform (including some caps), you can be sure he or she is an officer with the pay-grade of O-1.

Strangely, the military puts silver above gold in its rank devices. So even though an O-2 (lieutenant junior grade or first lieutenant, depending upon the service) outranks an O-1, the collar device of the O-2 is silver, whereas

---

*For a better understanding of the *actual* differences between officers and enlisted, you should refer to "Military Titles," chapter 1.

that of the O-1 is gold. This occurs again at the O-4 (lieutenant commander or major) and O-5 levels, where the commander or lieutenant colonel wears a silver oak leaf and the more junior O-4 wears a gold one.

Although I have been referring to these as collar devices, they are also worn on the shoulders at various times and on some caps or hats. The size of these devices will vary accordingly.

As you can see on page 74, Navy and Coast Guard officers have some additional ways of wearing their ranks. Gold stripes are often worn by officers of these two services, either on their lower sleeves or on things called "shoulder boards." As you can see, the various combinations of stripes determine their actual ranks. An ensign (O-1) wears a single half-inch stripe, a lieutenant junior grade (O-2) wears a half-inch plus a quarter-inch stripe, a lieutenant (O-3) wears two half-inch stripes, and so on. When these officers reach flag rank (become admirals), they keep going with the stripes on their sleeves (using combinations of one-inch and half-inch stripes), but on their shoulder boards (now gold instead of black) they wear one or more silver stars along with an anchor to show their ranks (a vice admiral having three stars, for example). You will note that there is some logic in that the number of stars on their shoulder boards matches the number of stars they wear when using collar devices.

Warrant officers in the Navy, Coast Guard, Marine Corps, and Army have their own rank devices as indicated on page 75.*

Covers

Another way to sometimes tell an officer from an enlisted person is by his or her cover ("cap" or "hat" in civilian lingo); although this can get a bit complicated as you will see. As mentioned above, the same things we called "collar devices" also sometimes appear on covers.

All officers of all services wear (at some time or another) what we call a "combination cover." These are the caps that have a visor on the front and are similar to what airline pilots, bus drivers, and police officers often wear. All enlisted personnel in the Army, Air Force, Marines, and Coast Guard also sometimes wear combination covers, and in the Navy, paygrades E-7 through E-9 also wear them.

---

*There are no warrant officers in the Air Force.

One component of these combination covers is the chin strap, a narrow strap that is just above the visor (called a "chin strap" because it can be extended to hook under the chin in windy conditions, though it rarely is). If that chin strap is gold, the person wearing it is an officer. In the Army, Navy, and Coast Guard, these chin straps are always gold for officers and always black for enlisted personnel. In the Air Force, they are black for both officers and enlisted. In the Marine Corps, they are gold (with a thin red stripe) for officers and black for enlisted when they are wearing a white cover, but are black for both officers and enlisted when wearing the green combination cover. (Confused yet?)

These covers are worn by male officers, male chiefs (enlisted), and female officers, respectively.

If the visor of a combination cover has "stuff" on it—what we usually refer to informally as "scrambled eggs"—the person is definitely an officer—a more senior one. In all the services except the Air Force, these added decorations are gold combinations of oak leaves and acorns; in the Air Force, they are silver clouds and lightning bolts. Just in case you were getting the hang of this, I will point out that in the Army, Air Force, and Marines, an officer gets to wear "scrambled eggs" on his or her cover at the O-4 level and above; in the Navy and Coast Guard, they are worn by O-5 and above!*

Combination covers for women are designed differently and vary according to service. In the Navy, women officer's covers have a gold chin strap just like the men's, but only for O-4s and below. Commanders and above do not have gold chin straps but do have "scrambled eggs"; these are mounted around the head band rather than on the visor, however.

Keep in mind that officers do not always wear combination covers. You will also see officers wearing garrison caps (also called "fore-and-aft caps" in the Navy and Coast Guard), various camouflage caps, baseball-type caps, berets, and helmets. Sometimes these will have an indication of rank on them, sometimes not.

---

*When I was serving as operations officer in a destroyer in the western Pacific, I received an incoming message that my executive officer had been promoted to commander (O-5); to deliver the good news, I was able to "steal" his combination cover from his stateroom and put it as his place at the breakfast table with an array of freshly scrambled eggs piled onto the visor.

Other Clues

An individual wearing white shoes is an officer or an E-7, E-8, or E-9 in the Navy, Marine Corps, or Coast Guard.

If you see someone wearing green trousers or skirt with a black stripe running down the whole length, that individual is an Army officer.

If you see someone wearing white shoes and a gold chin strap on the cap, you can assume he or she is an officer in the Marines, Navy, or Coast Guard. If the individual is wearing white shoes and a black chin strap, he or she is an E-7, E-8, or E-9 in the Navy.

Men wearing "choker" uniforms in white are often officers in the Navy, Marines, or Coast Guard, but be aware that E-7s through E-9s in the Navy also wear choker whites.

An individual wearing khaki from head to toe is an officer or an E-7 through E-9 in the Navy. The collar devices will tell you the difference.

As stated above, if you see someone wearing a gold chin strap, that person is an officer.

Staff versus Line Officers in the Navy

The majority of naval officers are *line* officers, whereas others are *staff* officers. The distinction is a little complicated, but essentially, line officers are those who are eligible for command at sea (of ships, submarines, or aviation units), and staff officers are those with specialty occupations like doctors, lawyers, or chaplains.

Line officers all look alike in that, depending upon the uniform they are wearing, they all wear their rank devices on both collars and they all wear a gold star with the stripes on their lower sleeves or on their shoulder boards.

Staff officers, on the other hand, have distinguishing devices that they wear on their uniforms to tell you what their specialties are (see page 76). If they are wearing collar devices, these devices are worn on their left collars whereas their rank devices are on their right collars (memory aid: "Rank on Right"). If they are wearing shoulder boards or stripes on their lower sleeves, the star worn by line officers is replaced by this same specialty device.

These devices vary according to the specialty. As you become familiar with them, you will be able to tell something about these individuals without a word spoken between you. For example, if you encounter an officer with a cross on her left collar point, you can be sure she is a Christian chaplain. There are other

Navy, Army, Air Force, Marine Corps, and Coast Guard Officer Ranks

Captain/Colonel
O-6

Commander/
Lieutenant Colonel
O-5

Lieutenant Commander/Major
O-4

Lieutenant/Captain
O-3

Lieutenant Junior Grade/
First Lieutenant
O-2

Ensign/Second Lieutenant
O-1

Fleet Admiral (Navy)/
General of the Air Force/Army
O-10

Admiral/General
O-10

Vice Admiral/
Lieutenant General
O-9

Rear Admiral (upper half)/
Major general
O-8

Rear Admiral (lower half)/
Brigadier General
O-7

Navy and Coast Guard Officer Ranks

Shoulder	Sleeve		Shoulder	Sleeve

Captain
O-6

Commander
O-5

Lieutenant Commander
O-4

Lieutenant
O-3

Lieutenant Junior Grade
O-2

Ensign
O-1

Admiral/Commandant of the Coast Guard
O-10

Vice Admiral
O-9

Rear Admiral (Upper Half)
O-8

Rear Admiral (Lower Half)
O-7

Note: Coast Guard background is lighter blue.

Navy, Coast Guard, Marine Corps, and Army Warrant Officer Ranks*

	Navy & Coast Guard Shoulder	Collar	Marine Corps Collar	Army Collar
Chief Warrant Officer 5 W-5				
Chief Warrant Officer 4 W-4				
Chief Warrant Officer 3 W-3				
Chief Warrant Officer 2 W-2				
Chief Warrant Officer 1 W-1				

*No Warrant Officer Ranks in the Air Force

Officer Staff Corps Devices

Chaplain Corps Symbols

Doctor   Nurse   Dentist   Medical Service
                                                Corps

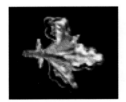

Supply Corps   Lawyer   Civil
                                       Engineering
                                       Corps

Navy and Coast Guard Enlisted Ranks

Petty Officer First Class
E-6

Petty Officer Second Class
E-5

Petty Officer Third Class
E-4

Seaman
E-3

Seaman Apprentice
E-2

Seaman Recruit
E-1

Master Chief Petty Officer of
the Navy/Coast Guard
E-9

Master Chief Petty Officer,
Fleet/Force/Command
Master Chief Petty Officer
E-9

Senior Chief Petty Officer
E-8

Chief Petty Officer
E-7

Marine Corps Enlisted Ranks

Staff Sergeant
E-6

Sergeant
E-5

Corporal
E-4

Lamce Corporal
E-3

Private First Class
E-2

Private
E-1

Sergeant Major
of the Marine Corps
E-9

Master Gunnery Sergeant
Sergeant Major
E-9

Master Sergeant
First Sergeant
E-8

Gunnery Sergeant
E-7

Army Enlisted Ranks

Staff Sergeant
E-6

Sergeant
E-5

Corporal, Specialist
E-4

Private First Class
E-3

Private First Class
E-2

Private
E-1

Sergeant Major
of the Army
E-9

Sergeant Major,
Command Sergeant Major
E-9

Master Sergeant
First Sergeant
E-8

Sergeant First Class
E-7

Air Force Enlisted Ranks

Technical Sergeant
E-6

Staff Sergeant
E-5

Senioir Airman
E-4

Airman First Class
E-3

Airman
E-2

Airman Basic
E-1

Chief Master Sergeant of the
Air Force
E-9

Chief Master Sergeant, First Sergeant,
Command Chief Master Sergeant
E-9

Senior Master Sergeant,
First Sergeant
E-8

Master Sergeant,
First Sergeant
E-7

Armed Forces Ribbons and Medals

Medal of Honor

Navy Cross

Defense Distinguished
Service Medal

Distinguished Service
Medal

Silver Star

Defense Superior
Service Medal

Legion of Merit

Distinguished Flying
Cross

Navy/Marine Corps
Medal

Bronze Star

Purple Heart

Defense Meritorious
ServiceMedal

Meritorious Service
Medal

Air Medal

Joint Service
Commendation Medal

Navy Marine Corps
Achievement medal

Combat Action Ribbon

Presidential Unit
Citation

Joint Meritorious Unit
Award

Navy Unit
Commendation

Meritorious Unit
Commendation

Navy "E" Ribbon

POW Medal

Good Conduct Medal

Navy Reserve
Meritorious Service
Medal

Fleet Marine Force
Ribbon

Navy Expeditionary
Medal

Armed Forces Ribbons and Medals (continued)

National Defense
Service Medal

Korean Service Medal

Antarctic Service Medal

Armed Forces
Expeditionary Medal

Vietnam Service Medal

Southwest Asia Service
Medal

Kosovo Campaign
Medal

Armed Forces Service
Medal

Humanitarian Service
Medal

Sea Service Deployment
Ribbon

Navy Arctic Service
Ribbon

Navy Reserve Sea
Service Ribbon

Navy/Marine Corps
Overseas Service
Ribbon

Navy Recruiting Service
Ribbon

Navy Recruit Training
Service Medal

Armed Forces Reserve
Medal

Navy Reserve Medal

Philippine Presidential
Unit Citation

Republic of Korea
Presidential Unit
Citation

Republic of Vietnam
Presidential Unit
Citation

Republic of Vietnam
Gallantry Cross Unit
Citation

Armed Forces Ribbons and Medals (continued)

Republic of Vietnam
Civil Actions Unit
Citation

United Nations Service
Medal

United Nations Medal

NATO Medal

Multinational Force and
Observers Medal

Inter American Defense
Board Medal

Republic of Vietnam
Campaign Medal

Kuwait Liberation
Medal (Kingdom of
Saudi Arabia)

Kuwait Liberation
(Kuwait)

Rifle Marksmanship
Medal

Pistol Marksmanship
Medal

Devices

GOLD STAR
Denotes subsequent
awards of the same Navy
decoration

SILVER STAR
Worn in lieu of five gold
stars

BRONZE STAR
Represents participation in
campigns or operations,
multiple qualification or an
additional award to any of
the various ribbons on
which it is authorized. Also
worn to denote first award
of the single-mission Air
Medal after Nov. 22,
1989.

SILVER SERVICE STAR
Worn in lieu of five bronze
stars

BRONZE OAK LEAF
CLUSTER
Represents second and
subsequent entitlements of
awards

Devices (continued)

SILVER OAK LEAF
CLUSTER
Worn for the 6th, 11th, or
in lieu of five bronze oak
leaf clusters

"V DEVICE"
Authorized for acts or
service involving direct
participation in combat
operations

EUROPE AND ASIA
CLASPS
Worn on the suspension
ribbon of the Navy
Occupation Service Medal

SILVER "E"
Denotes Expert Marksman
qualification

BATTLE "E" DEVICE

FLEET MARINE FORCE
COMBAT OPERATIONS
INSIGNIA
For Navy personnel
attached to Fleet Marine
force units participating in
combat operations

"M" DEVICE
Denotes Naval Reserve
mobiilzation in support of
certain operations

"E" DEVICE
Denotes four or more
Battle "E" Awards

KUWAIT LIBERATION
CLUSTER

BRONZE "S"
Denotes Sharpshooter
Marksman qualification

STRIKE/FLIGHT DEVICE
Bronze Arabic numeral
denotes the total number
of strike/flight awards of
the Air Medal earned
subsequent to April 9,
1952

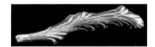

3/16 PALM
Worn on the Republic of
Vietnam Gallantry Cross
Unit Citation and Republic
of Vietnam Civil Actions
Unit Citation ribbons

REPUBLIC OF VIETNAM
CAMPAIGN CLASP

Pins and Badges Worn by Armed Forces Personnel

Astronaut

Naval Astronaut (NFO)

Naval Aviator

Naval Aviation Observer and Flight Meteorologist

Flight Surgeon

Flight Nurse

Naval Flight Officer (NFO)

Aviation Experimental Psychologist and Aviation Physiologist

Enlisted Aviation Warfare Specialist

Naval Aviation Supply Corps

Aircrew

Combat Aircrew

Speical Warfare (SEAL)

Special Operations

Special Warfare Combatant Craft Crewman

Surface Warfare Officer

Enlisted Surface Warfare Specialist

Surface Warfare Nurse Corps

Surface Warfare Medical Corps

Surface Warfare Dental Corps

Surface Warfare Medical Service Corps

Surface Supply Corps

Submarine (officer)

Submarine (enlisted)

Pins and Badges Worn by Armed Forces Personnel (continued)

Submarine Medical

Submarine Engineering
Duty

Submarine Supply Corps

Submarine Combat Patrol

SSBN Deterrent Patrol
(20 Patrols)

SSBN Deterrent patrol

Seabee Combat Warfare
Specialist (officer)

Seabee Combat Warfare
Specialist (enlisted)

Naval Parachutist

Basic Parachutist

Naval Reserve Merchant
Marine

Integrated Undersea
Surveillance System
(officer)

Integrated Undersea
Surveillance System
(enlisted)

Master Explosive Ordnance
Disposal Warfare

Senior Explosive Ordnance
Disposal Warfare

Basic Explosive Ordnance
Disposal Warfare

Fleet Marine Force (FMF)
Enlisted Warfare Specialist

Master Diver

Pins and Badges Worn by Armed Forces Personnel (continued)

Diving Medical Officer

Diving Officer

Diver (medical technician)

First Class Diver

Second Class Diver

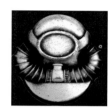

Scuba Diver

Deep Submergence
(enlisted)

Deep Submergence
(officer)

Presidential Service Badge

Vice Presidential Service
Badge

Office of the Secretary of
Defense

Joint Chiefs of Staff

Recruiting Command for
Exellence

Recruiter

Career Counselor

Pins and Badges Worn by Armed Forces Personnel (continued)

Division Commander for Excellence

Division Commander

Command at Sea

Command Ashore/Project Manager

Small Craft (officer)

Small Craft (enlisted)

Craftmaster

Master Chief Petty Officer of the Navy

Fleet Master Chief Petty Officer

Force Master Chief Petty Officer

Command Master Chief Petty Officer

U.S. Navy Police (officer)

appropriate symbols for other religions, such as tablets with a Star of David for Jewish chaplains and a crescent for Muslim chaplains.

An officer wearing a gold oak leaf with a silver acorn centered on it is a doctor; one wearing a gold oak leaf but no acorn is a nurse;* one wearing a gold oak leaf and two silver acorns at the bottom is a dentist; and one wearing a gold oak leaf sprouting from a gold twig is a medical service officer.

Supply Corps officers wear a differently shaped oak leaf with three gold acorns that is traditionally (and unofficially) called a "pork chop," and lawyers wear a strange-looking device that consists of two curved gold oak leaves bracketing a silver "mill rinde" (an object that once kept stone-grinding wheels an equal distance apart and was used by sixteenth century English lawyers to symbolize the "wheels of justice grinding exceedingly fine and even").

Civil engineers wear a cluster of overlapping, elongated gold oak leaves with two silver oak leaves centrally positioned.

Warrant officers have their own devices reflecting their specialties, such as crossed quills for "Ship's Clerk" and a three-bladed propeller for an "Engineering Technician." There are too many of these to cover here, but you are less likely to encounter these than the ones worn by line and staff corps officers.

Enlisted Personnel

Because there are far more enlisted personnel in the armed services than there are officers, you are mathematically more likely to encounter the former than the latter. (But the proportions can change significantly, depending upon where you work—if you are in the E-Ring of the Pentagon, for example, you will see *many* officers!)

As already stated, if you see someone wearing the uniform that is unofficially called "Crackerjack," that person is an enlisted Sailor in the Navy (E-6 or below).

Note that in the Navy—unlike all the other services—the uniforms of E-7s through E-9s (chief petty officers, senior chief petty officers, and master chief petty officers) make a marked change, so that these individuals look

---

*Note that a nurse's oak leaf is shaped differently than the oak leaf rank device of a lieutenant commander.

more like officers in the Navy than they do other (E-6 and below) enlisted personnel.

Just as with officers, the most reliable way to tell enlisted personnel is by their rank devices. Familiarizing yourself with the various indications of enlisted rank on pages 77–80 is virtually foolproof but not easily done, because there are so many (there is much less commonality among the enlisted rank devices of the various services than among the officers).

One clue is that if someone is wearing an indication of rank on his or her *upper* arm, that person is enlisted. This is true of all services: though enlisted personnel may wear collar devices with some uniforms, you will never see officers wearing any indication of rank on their upper arms. In the other services these are worn on both arms, but in the Navy and Coast Guard, they are worn only on the left arm.

Referring to the enlisted rank chart on page 77, you can see that Navy enlisted personnel above E-3 wear a patch on their arms that is a combination of an eagle (more traditionally called a "crow") and one or more chevrons (a chevron looks like a flattened "V" and there is one for a petty officer third class, two for a second class, etc.).

In the Navy and Coast Guard, enlisted personnel wear one or more long, diagonal stripes on the *lower* left forearm of some of their uniforms (do not confuse these with the smaller diagonal stripes worn on the *upper* arm by junior enlisted personnel in both services to indicate paygrade); officers do not wear these service stripes. These indicate years of service: one stripe for every four years of completed service. So a person wearing three of these stripes has *at least* twelve years of service—but they might have as much as fifteen years, eleven months, and thirty days of service because they do not rate another stripe until another four years of service have been *completed*. In the Navy, these stripes will be dark blue on white uniforms, but they can be either red or gold on blue uniforms (the difference is these stripes go from red to gold after twelve continuous years of service with good conduct).

RATING SYMBOLS

The Navy and Coast Guard are unique among the services in that enlisted personnel wear indications of their occupational specialties (called "ratings" in the Navy) on their uniforms.

If you look at the rank chart (page 77) for Navy enlisted personnel, you will see a pair of crossed anchors between the chevrons and the "crow." These are not the same for every petty officer. The crossed anchor symbol in this example represents the occupational specialty known as "boatswain's mate," one who is a specialist at seamanship, working with boats, anchoring equipment, mooring lines, and so on. If an individual has a different rating, he or she wears some other symbol in place of the crossed anchors. For example, a Hospital Corpsman would have a medical caduceus instead. A Gunner's Mate wears a pair of crossed cannons, and a Musician is represented by a lyre.

With some of these symbols, there are significant clues as to the Sailor's occupational specialty, as with the caduceus and the lyre. Others are less helpful (I doubt you would have realized that those were crossed cannons if I had not told you). Some of these rating symbols are downright mysterious, such as the one for a Photographer's Mate (described in Uniform Regulations as a "winged graphic solution of photographic problem") or the globe that represents an Electrician's Mate.

As you can see in figure 3.1, there are too many of these to memorize unless you are a trivia nut, but some familiarity with them goes a long way in knowing what an individual's primary job is in the Navy.

RIBBONS AND MEDALS

In the military, special achievements are recognized by the awarding of ribbons and medals. You are probably familiar with the Medal of Honor, the highest honor that is awarded for combat valor, and you may well have heard of the Purple Heart, the medal that is awarded to those who have been wounded in combat.

When a person is awarded one of these medals, or one of the many others that have been authorized, she or he only wears the actual medal on special occasions, such as for a formal change of command or a visit by a dignitary. More frequently, an abbreviated version is worn on dress uniforms (not working ones). That version is called a "ribbon" and is created by taking a portion of the cloth from which the medal is suspended. If you look at the ribbons that represent that Medal of Honor and the Purple Heart, you will see the similarity.

AB
Aviation Boatwain's Mate
ABE—Equipment
ABF—Fuel
ABH—Handling

AC
Air Traffic Controller

AD
Aviation Machinist's Mate

AE
Aviation Electrician's Mate

AG
Aerographer's Mate

AM
Aviation Structural
Mechanic
AME—Equipment

AO
Aviation Ordnanceman

AS
Aviation Support
Equipment Technician

AT
Aviation Electronics
Technician

AW
Aviation Warfare Ststems
Operator

AZ
Aviation Maintenacnce
Administrationman

BM
Boatswain's Mate

BU
Builder

CE
Construction Electrician

CM
Construction Mechanic

Figure 3.1 Navy Ratings

CT
Cryptologic Technician
CTA—Administrative
CTI—Interpretive
CTM—Maintenance
CTO—Communications
CTR—Collection
CTT—Technical

DC
Damage Controlman

EA
Engineering Aide

EM
Electrician's Mate

EN
Engineman

EO
Equipment Operator

ET
Electronics Technician

FC
Fire Controlman

FT
Fire Control Technician

GM
Gunner's Mate

GS
Gas Turbine System
Technician
GSE—Electrical
GSM—Mechanical

HM
Hospital Corpsman

HT
Hull Maintenance
Technician

IC
Interior Communications
Electrician

Figure 3.1 *(continued)*

IS
Intelligence Specialist

IT
Information Systems
Technician

JO
Journalist

LI
LIthographer

LN
Legalman

MA
Master-at-Arms

MM
Machinist's Mate

MN
Mineman

MR
Machinery Repairman

MT
Missile Technician

MU
Musician

NC
Navy Counselor

OS
Operations Specialist

PC
Postal Clerk

PH
Photographer

PN
Personnelman

PR
Aircrew Survival
Equipmentman

QM
Quartermaster

Figure 3.1 *(continued)*

RP
Religious Program
Specialist

SH
Ship's Serviceman

SK
Storekeeper

ST
Sonal Technician
STG—Surface
STS—Submarine

SW
Steelworker

TM
Torpedoman's Mate

UT
Utilitiesman

YN
Yeoman

Figure 3.1 *(continued)*

You will rarely see the actual medals, so we will concentrate on the ribbons, because these are worn on a regular basis.

Also, you should know that every medal has an abbreviated ribbon, but there are a number of ribbons that have been created that do *not* have an associated medal, for example, the Combat Action Ribbon and the Battle Efficiency "E" (also called the "Navy E").

Most of the medals and ribbons are awarded to individuals, but some are awarded to whole units at a time. For example, a ship that excels during a battle or a major operation might be awarded the Presidential Unit Citation, the Navy Unit Citation, or the Meritorious Unit Commendation. Virtually any unit (aircraft squadrons, SEAL teams, construction battalions, schools, systems commands, etc.) can be given a unit award for outstanding service. In the sea services (Navy, Marine Corps, Coast Guard), when a unit award is received, all personnel assigned to that unit at the time of the special service is authorized to wear that unit award. These individuals then wear that ribbon for the rest of their careers. The Army does this very differently. Individual Soldiers wear unit awards only when they are serving in a unit that received the award sometime in its past (regardless of whether the individual was in the unit at the time); they then remove the ribbon from their uniforms when they are transferred out of that unit. One other distinction is that they wear these unit awards on the right side of their chests, separate from the other ribbons, which are worn on the left. Sea service personnel integrate their unit ribbons among their others.

When the unit receiving an award includes civilians, a special lapel pin version of the ribbon is awarded to them. Civilians are eligible for some other awards as well—such as the Navy Distinguished Civilian Service Award and the Navy Distinguished Achievement Science Award.

If you look at pages 80–82, you will see that, like ratings, there are too many ribbons for all but the very gifted to memorize. But recognizing some of these allows you to know certain things about an individual just by looking at these patterns of color. For example, the Combat Action Ribbon tells you that a person was in some kind of combat, and if you recognize a Purple Heart among an individual's ribbons, you know that he or she was wounded in combat. The Navy Cross is second only to the Medal of Honor for heroism in combat, and the Navy and Marine Corps Medal is given for heroism not involving combat (such as rescuing someone from a fire). Someone wearing the Vietnam Service Medal is obviously getting on in years, and you probably would not impress someone wearing the Antarctica Service Medal by saying, "It sure is cold out there today."

Note that the ribbons on pages 80–82 are arranged in their order of precedence. When wearing them on a uniform, a Sailor who has been awarded more than one award will arrange them in rows of three in the order they appear in this figure. Higher precedence awards go above and to the right of lower ones (appearing to the left to an observer). So someone who had been awarded the Navy and Marine Corps Commendation Medal, the Southwest Asia Service Medal, and the Kuwait Liberation Medal (awarded by the Saudi Arabian government) would wear them in that order.

As more awards are earned, they are inserted in the proper order, adding rows of one, two, or three on top.*

To simplify things, individuals have the option of wearing only the top three ribbons in a single row; so if you see someone wearing only three ribbons and one of them is a high precedence award like the Silver Star or the Legion of Merit, chances are that individual has a bunch more ribbons sitting on the dresser at home.

You will see various attachments on some of the ribbons. Some of these (like the palm tree and crossed swords on the ribbon awarded by the Saudi government for the liberation of Kuwait) are a part of the ribbon and are always there. Other devices are added to ribbons to represent such things as multiple campaigns in the same theater of operations or the number of sorties flown. A "V" (for "valor") indicates the award was received for service under combat conditions, and an "O" indicates that the award was given under operational ("hands-on") conditions (as opposed to administrative circumstances). When individuals earn an award more than once, they do not wear the same ribbon twice; stars are added instead. So if you see someone wearing a Bronze Star ribbon with a gold star on it, you know she or he has received it on two separate occasions.

---

*Soldiers, Airmen, and Marines who have a *lot* of ribbons wear them in rows of *four*, but Navy awards are always in rows of three, no matter how many a Sailor has been awarded.

PINS AND BADGES

Another thing common to uniforms are the various pins and badges that indicate special qualifications or assignments. A quick scan of pages 85–88 will tell you that there are a lot of them. A closer look will tell you that there are some real Sherlock Holmes clues here if you care to learn some of them.

Warfare Qualifications

Most of us are familiar with the concept of aviator wings. There is a certain logic that goes with a person who can fly airplanes wearing a set of wings on his or her chest. We have seen airline pilots wearing them and aviators in the movies doing the same.

When submarines appeared, it seemed similarly logical to recognize the submariner in the same manner. Like the aviators who had taken to the sky like birds, these pioneers of a new realm, descending to the depths like fish, also deserved to proudly proclaim their special abilities to the world by wearing a pair of stylized dolphins on their chests. There was probably some thought given to the idea that if we were going to ask people to do such dangerous things, it might help to let them "boast" a bit by wearing their daring for all the world to see.

This seemed to work for a while, until one day, someone realized that this practice was not quite fair. Surface Sailors had been venturing into the hostile realm of the sea for thousands of years, taking risks and using specially acquired skills in a manner that is comparable to that of their fellow "airdales" and "bubbleheads." But because they had been doing it for so long, it was taken for granted, and they were not permitted to "boast" about it on their chests. This was made worse by the realization that many people in Navy uniforms did not have the same knowledge or skills as these mariner-warriors (like doctors, lawyers, chaplains, etc.)—all of these looked alike (in the chest area at least), whereas the aviators and submariners looked special.

So the "surface warrior" was born. This new (actually quite old) breed of Sailor was at last given due recognition by being permitted to add a special pin to his or her chest that shows the bow of a ship, a stylized series of waves, and a pair of crossed swords for officers or cutlasses for enlisted surface warriors. You might hear this pin called (unofficially, of course) "water wings."

The inevitable occurred, and soon pins were designed for SEALS, SeaBees, Sailors assigned to Marine Corps units (Fleet Marine Force Enlisted Warfare Specialists), and others with special warfare qualifications.

Aviation Variations

At first, only pilots wore wings. But over time, others making special contributions to aviation were given their own wings. Officers who controlled weapons, navigation, and other systems were distinguished from pilots (officially called "Naval Aviators") by a new designation called "Naval Flight Officer" (or more commonly, "NFO"). Pilots wear wings with a single vertically oriented anchor behind a federal shield; NFOs wear wings with a pair of crossed anchors behind the shield.

Astronauts, Aircrew, Flight Surgeons, and others also wear specialized wings. Enlisted personnel who complete a rigorous qualification program, similar to those in submarines and surface ships are awarded the silver wings of an "Enlisted Aviation Warfare Specialist."

Combinations

"Flight Surgeons" are doctors with special qualifications for handling the specialized medical needs of aviators. Others with special combinations of knowledge and skills are also recognized by specialized pins. Supply officers on surface ships, engineers on submarines, nurses on surface ships, and a host of other combinations all have specially designed pins proclaiming their special contributions to naval service. These look much like their parent warfare pins but have different features, like the "Naval Aviation Supply Corps" officer's pin, which has a supply oak leaf ("pork chop") in place of the shield and anchors of the pilots and NFOs.

Other Chest Hardware

Besides those pins that proclaim a warfare specialty, there are other pins and badges that recognize additional qualifications or special assignments.

Special Qualifications

Individuals who jump out of a perfectly good airplanes wear "jump wings" if they (a) wear a parachute and (b) do it as part of their Navy duties or training. Those who complete jump school training wear the silver version of jump wings, whereas those individuals who have completed ten or more authorized jumps earn the gold "Naval Parachutist" wings.

There are other pins that also recognize other special qualifications. You can see on pages 86–87 that there are a number of pins for divers and several for those who disarm bombs and the like ("Explosive Ordnance Disposal").

Command Pins

An officer who has been given command of a ship, submarine, or an aviation unit (such as a squadron) wears the "Command-at-Sea" pin, a circular pin with a star as its most prominent feature. An officer who commands a unit ashore (such as a naval base) or heads a large, important project (like the development of major new weapons system) wears a similar pin that has a trident as its predominant feature.

Where these pins are worn gives you an additional clue to "read." If you see one of these pins worn on the right side of the chest, that individual is currently in command. If the pin is on the left side, that individual was once in command but is not currently. Once an individual is promoted to flag rank, he or she no longer wears command pins.

Command Pin Equivalents

During the Vietnam War, when junior officer and senior enlisted Sailors were in charge of coastal and riverine craft, engaging the enemy "up close and personal"—something approximating the responsibility of command of ships at sea—the decision was made that they too should have a pin. The "Small Craft" pin was adopted (in a gold version for officers and a silver version for enlisted).

For those in charge of small craft such as tugs, there is a "Craftmaster" badge.

Command Master Chiefs and the Like

Recognizing the value of the experience and accumulated wisdom of senior enlisted personnel, the Navy appoints one individual at nearly every command to serve at the right hand of the commanding officer, providing advice on matters affecting the enlisted personnel in the command and sharing in some of the decisions that affect the command as a whole. At most commands these individuals are known as the "Command Master Chief." Aboard submarines they are known as the "Chief of the Boat."

Those who are performing similar functions, serving admirals rather than less-senior commanding officers, are known by titles reflecting their particular assignments. An individual serving as the principal enlisted advisor to a fleet commander is known as the "Fleet Master Chief." One serving a force commander or a flag officer of a major command (like the Naval Education and Training Command) is known as a "Force Master Chief."

All of these individuals wear special badges as pictured on page 88. Not all of the variations are shown, but they all essentially look much alike.

West Wing and the Like

Individuals who are assigned very special (and prestigious) billets wear specially designed badges. These are:

Presidential Service Badge
Vice Presidential Service Badge
Office of Secretary of Defense
Joint Chiefs of Staff

The Presidential Service Badge and Vice Presidential Service Badge are worn on the right of the uniform and the other two are worn on the left, below the ribbons. These are worn both during and after the actual assignment, so if you see someone wearing one of these badges, you will know that they are either currently in that billet or served in it at some time in their past. Either way, it is a great conversation starter.

Key Personnel Billets

Because people are the Navy's most important resource, there are several assignments that are recognized by special identifying badges. Individuals serving as recruiters for the Navy, Recruit Division Commanders (training recruits at Boot Camp—equivalent to Marine Drill Instructors), or Career Counselors (specially trained to assist Sailors in making important career decisions, such as reenlisting) all wear specially designed emblems on their uniforms (left side, below the ribbons). These badges are worn only while actually serving in the billet.

Security and Law Enforcement

Just as police officers wear special badges in civilian life, naval personnel who are providing security or law enforcement functions wear special badges on their uniforms. Examples of such personnel are those providing base patrol or serving as brig (jail) and gate guards. Aboard ship, individuals who perform these functions are often called "Master-at-Arms." These badges are only worn by personnel actually performing one of these functions.

Silver versus Gold

Be careful *not* to apply logic to the meaning of silver versus gold in these various uniform accoutrements. We have seen that silver outranks gold among officer's collar devices, yet some (not all) of the warfare pins come in both silver and gold varieties—and it is the officers who wear the gold and the enlisted get the silver! But command master chief pins (recognizing senior enlisted personnel) are predominantly gold (although they contain both silver and gold). Some of the pins (like basic jump wings and explosive ordnance disposal pins) come in silver only and are worn by both officers and enlisted.

AIGUILETTES

You may on occasion see someone wearing a colored ropelike loop (or loops) over his or her shoulder. Sometimes these are rather elaborate, with multiple loops and some metal hardware (called "pencils") mixed in.

Some are less elaborate, having only a single loop or a few loops perched on the end of the shoulder and going under the arm.

These are called "aiguillettes" and have special meanings. They are worn by various aides to high-ranking officers (admirals and generals) and officials (such as the President, Vice President, and Secretary of Defense). They are also worn by naval attachés at diplomatic missions.

Special aiguillettes are worn by members of drill teams and by various ranking individuals in special training commands, such as Recruit Division Commanders at Recruit Training Command (Boot Camp).

BRASSARDS

Brassards are bands of cloth, suitably marked with symbols, letters, or words, indicating a *temporary* duty to which the wearer is assigned, such as Officer of the Day (OOD), Junior Officer of the Day (JOOD), Master-at-Arms (MAA), or Shore Patrol (SP). They are worn on the right arm, midway between the shoulder and the elbow, on outermost garments.

Another variation is the mourning badge, made of black crepe, which may be worn on the sleeve of the outermost garment, halfway between shoulder and elbow for funerals, memorial services, and similar occasions. Officers wear it on the left sleeve and enlisted personnel wear it on the right.

FINAL THOUGHTS

Reading uniforms is not the easiest thing in the world. Trying to memorize all of the variations is probably not worth your time. A better approach is to be generally familiar with what these things mean, commit a few of the more prominent features to memory, and keep this guide handy for reference so that you can identify specific variations as you encounter them. You can probably get by without knowing them, but being able to pick the senior officer out in a crowded room might be a career enhancing piece of knowledge. Also, you can equip yourself for that awkward moment when you meet someone for the first time and wonder, "What do I talk about?" Trust me, you will score a lot more points with someone in uniform if you say, "I see you served in the first Gulf War," rather than, "I hear it's supposed to rain later today."

QUICKREFS

See pages 73–80 and Figure 3.1 in this chapter for quick reference to many uniform items:

Clues to different service uniforms:

- Navy personnel wear "Crackerjacks," all white, all black, all khaki, or khaki and black, depending upon the season, rank, and circumstances.
- "Choker" whites mean Navy, Coast Guard, or Marine Corps. "Choker" blues mean Marine Corps only (except at service academies).
- Light green shirts are Army. Light blue shirts are Air Force or Coast Guard. White shirts are Navy. Khaki shirts are Navy or Marine Corps.
- Khaki neckties are Marine Corps. Blue neckties are Air Force or Coast Guard. Black neckties are Army or Navy.

- Camouflage uniforms are worn by all services and are difficult to tell apart, but the name of the service usually appears over one of the pockets.

Identifying officers versus enlisted personnel:
- The most reliable way is to familiarize yourself with the various rank devices worn on uniforms (see pages 73–75 and 77–80).
  —Ranks on the upper arms indicate enlisted personnel.
  —Collar devices are virtually the same for officers in all services.
  —Ranks on the lower sleeve or on large shoulder boards indicate Navy or Coast Guard officers.
- Combination covers (visored caps) have a thin strap just above the visor (called a "chinstrap"); if that strap is gold, the person is an officer. Black chinstraps are worn by enlisted personnel of all services, but Air Force officers also have black chin straps, as do Marine officers when wearing their green uniforms.
- "Scrambled eggs" (clusters of oak leaves and acorns) on a visor is a dead giveaway that the individual is an officer (a senior one—O-4 and above in all services except Navy and Coast Guard, which are O-5 and above).
- Khaki from head to toe is an officer or an E-7, E-8, or E-9 in the Navy. The collar devices will tell you the difference.
- The majority of officers in the Navy wear a gold star above the stripes on their sleeves or on their shoulder boards to indicate they are *line officers* (eligible for command of ships, squadrons, etc.). Other officers with specialties such as law and medicine are known as *staff officers* and wear other symbols on their sleeves and shoulder boards, as well as on their left collar points.
- Enlisted personnel in the Navy, E-6 and below, wear the distinctive "Crackerjack" uniform. E-7, E-8, and E-9 wear a uniform more like officers wear.
- Navy enlisted personnel above E-3 wear an eagle, one or more chevrons, and a rating indicator between them. There are many different rating indicators, and they tell you the person's occupational specialty (see figure 3.1).
- Ribbons and medals are awarded for various achievements or for service in a particular area (see pages 81–83). Every medal has a ribbon, but not every ribbon has an equivalent medal. Medals are worn only for special occasions, but ribbons are worn on various dress uniforms.

- Pins and badges indicate a variety of special qualifications (such as being a qualified pilot or a diver), or they may indicate that a person is (or has been) in command, or they may indicate some specialized service (such as in the White House or as a recruiter). There are a large number of these as indicated on pages 85–88.
- Aiguillettes are worn over the shoulder and indicate some special service, such as being a key member of an admiral's staff or being in charge of a recruit division at Boot Camp.
- Brassards are bands of cloth temporarily worn on the sleeve to indicate such things as being a member of the Shore Patrol.

CHAPTER FOUR

The Paper Navy

Though word processing, e-mail, and other electronic marvels have greatly benefited the Navy, there remains a number of special practices that originated in the days when paper was supreme and some means had to be developed to organize great volumes of it. Long before there were such things as search engines, the Standard Subject Identification Code (SSIC) system helped the Navy keep its reams organized. Without a scroll bar, organizing letters and directives into specific formats helped readers scan them quickly to find the parts they needed. Before firewalls and passwords, a system of protecting sensitive documents from prying eyes was developed.

Though these communications systems might be different if they were developed from scratch today, they remain in effect and are still useful. Their existence alone makes understanding them essential if one is to function efficiently in this strange world that is part nautical, part clerical, part bureaucratic, and all too mysterious to the uninitiated.

In this chapter, we will shed some light on how the Navy communicates through correspondence and other means. We will explore the strange world of the Navy Directives System, learn to decipher the SSIC system, grapple with military classification of information, and the like. We will attach some logic (not too much) to those strange numbers that often appear on documents and learn how to make good use of the Navy's odd formats and alien ways of doing things.

NAVY PUBLICATIONS

Though the Constitution, various treaties, and Congress supply the fundamental laws governing the Navy, they are really only broad outlines. The Navy has various publications and official directives setting forth specific procedures for the daily operation of the Navy Department and for the administration of personnel.

The Navy Directives System will be explained later in this chapter, but it is worthwhile to take a quick look at some of the basic publications used frequently in various quarters of the Navy. Depending upon your job, some or all of these may be essential to you.

Note that Navy publications are often referred to by their abbreviated names or some combination of the originator's abbreviated name and some numbers. Most are decipherable once you get the hang of it, but they can be confusing at first. Some of the more common abbreviations you might encounter on Navy publications are provided below.

BUMED	Bureau of Medicine and Surgery
BUPERS	Bureau of Naval Personnel
COMNAVCRUITCOM	Commander, Navy Recruiting Command
DOD	Department of Defense
GPO	Government Printing Office
JAG	Judge Advocate General
NAVAIR	Naval Air Systems Command
NAVCOMP	Comptroller of the Navy
NAVEDTRA	Navy Education and Training
NAVMAT	Naval Material Command
NAVMILPERSCOM	Navy Military Personnel Command
NAVSO	Executive Offices of the Secretary of the Navy (Navy Staff Office)
NAVSUP	Naval Supply Systems Command
NETC	Naval Education and Training Command
OPNAV	Office of the Chief of Naval Operations
SECNAV	Secretary of the Navy
VA	Department of Veterans Affairs

The following list is by no means all-inclusive, but it is a good start toward understanding the pulp foundation of the paper Navy. Note that some of these publications have strange combinations of letters and numbers following them in brackets (like OPNAVINST 3120.32 and JAGINST 5800.7); these will make more sense later. Some of these publications are "joint" ones that apply to all the armed services and others are Navy only. Joint publications are often supplemented by Navy publications that cover any differences or additions.

- U.S. Navy Regulations (NAVREGS) (or often referred to as "Navy Regs") outlines the organizational structure of the DON and sets out the principles and policies by which the Navy is governed. It is a surprisingly thin document and covers surprisingly little. Yet there are some essential nuggets that cannot be found anywhere else.
- Standard Organization and Regulations of the U.S. Navy (OPNAVINST 3120.32) sets forth regulations and guidance governing the conduct of all members of the U.S. Navy and sets the standards for the organization of naval units. Though not technically correct, to understand its function, you might think of this as a kind of addendum to Navy Regulations. This is a much bigger publication than NAVREGS with a lot more information.
- Standard Subject Identification Code (SSIC) Manual (SECNAV M-5210.2) is described in some detail later in this chapter. It provides a numerical coding system that is useful in the filing and identification of documents.
- Standard Navy Distribution List (SNDL) (OPNAVNOTE 5400) lays out the administrative chain of command and provides addresses for fleet units and shore activities.
- Naval Military Personnel Manual (MILPERSMAN) (NAVPERS 15560) is a very important publication that affects military personnel in many big ways—including application for various educational programs, transfers, discharges, and separations.
- Manual for Courts-Martial, United States (MCM) describes the types of courts-martial established by the Uniform Code of Military Justice (UCMJ), defines their jurisdiction, and prescribes their procedures. It also covers such matters as nonjudicial punishment (NJP), reviews of court-martial proceedings, new trials, and limitations on punishment. This manual applies to all the armed services.
- Manual of the Judge Advocate General (JAGMAN) (JAGINST 5800.7) covers legal and judicial matters that apply only to the naval service. Included among these are instructions regarding boards of investigation and examining boards—their composition, authority, and procedures.
- Joint Federal Travel Regulations (JFTR) is issued in three volumes; only the first volume deals with actual travel. JFTR interprets the laws and regulations concerning the manner in which transportation is furnished,

travel for family members, the transportation of household goods, reimbursement for travel expenses, and similar information.

- U.S. Navy Travel Instructions (NAVSO P-1459) amplifies the rules laid down in volume 1 of the JFTR.
- Department of Defense Military Pay and Allowance Entitlements Manual (DODPM) covers statutory provisions for entitlements, deductions, and collections on military pay and allowances.
- Navy Pay and Personnel Procedures Manual (PAYPERSMAN) (NAVSO P-3050) contains detailed information about the procedures of the military pay system for members of the Navy.
- Enlisted Transfer Manual (TRANSMAN) (NAVPERS 15909) is the official manual for the distribution and assignment of enlisted personnel; it supplements the MILPERSMAN.
- Navy and Marine Corps Awards Manual (SECNAVINST 1650.1), or simply "Awards Manual," is issued by the SECNAV for guidance in all matters pertaining to decorations, medals, and awards, including how they are worn.
- Manual of Advancement (BUPERSINST 1430.16) addresses the administration of the Navy advancement (promotion) system. It explains the basic policies outlined in MILPERSMAN on eligibility requirements for advancement; the preparation of forms; the ordering, custody, and disposition of Navy-wide exams; the administration of examinations for advancement; changes in rate or rating; and the procedures for actual advancement.

DECIPHERING THE NUMBERS

The Navy has long used numbers as a means of document identification. Unfortunately, there are some numbering systems that do not do much more than uniquely identify a document; for example, the Naval Education and Training Command uses a series of NAVEDTRA numbers that may have some logic to it, but in all my years in the Navy, I never detected any.

But the good news is that, more often than not, these numbers in themselves convey information in an abbreviated form that can be very useful if you know how to read them. The Navy's SSIC system is the best example and is used in many ways Navy-wide. It is used as a means of identifying and filing correspondence, messages, official directives, and various other documents. Understanding how SSIC numbers work will take you a long way toward a better understanding of what is going on in the paper world of the Navy.

Standard Subject Identification Code (SSIC) System

To understand this system and how it is used, you must begin with a publication identified, appropriately enough, as the SSIC Manual. Like most Navy official documents it is a good substitute for a sleeping pill if you read it late at night (or after lunch), but it is full of useful information that can help you navigate through the seas of paper you are likely to encounter at most Navy commands. It explains the SSIC system and includes a listing of the codes that the Navy uses to make documents identifiable for filing and research purposes. If you deal with Navy paper much, you should keep a copy of it nearby. If you do not deal with Navy paper much, I want your job.

The codes are four- or five-digit numbers that are linked to particular subjects*. The general divisions of these codes are as follows:

- 1000–1999: Military Personnel. Subjects relating solely to the administration of military personnel. (Civilian personnel subjects are included in the 12000 series. General personnel subjects relating to both civilian and military personnel are included in the 5000 series.)
- 2000–2999: Telecommunications. General communication matters and related systems and equipment.

---

*The SSIC Manual has the annoying habit of appending the word "records" to many of the subject descriptions. This is redundant, unnecessarily limiting, and often inaccurate, because documents carrying these numbers frequently cover things other than mere record keeping. For example, if you went to the Navy Electronic Directives System (NEDS) Web site (http://neds.daps.dla.mil/), you would find that the SECNAV has issued a directive entitled, "Actions Based on Unacceptable Performance" that establishes "Department of the Navy policy for effecting reduction in grade and removal of employees based solely on unacceptable performance." (Fun subject!) Obviously, this directive is about actions, not just records. I find that more often than not, the subject descriptions in the SSIC Manual make more sense if you drop the word "records."

- 3000–3999: Operations and Readiness. Subjects related to such matters as operational plans, fleet operations, operational training and readiness, warfare techniques, operational intelligence, research and development, and geophysical and hydrographic support.
- 4000–4999: Logistics. Topics related to the logistical support of the Navy and Marine Corps, including procurement, supply control, property redistribution and disposal, travel and transportation, maintenance, construction and conversion, production and mobilization planning, and foreign military assistance.
- 5000–5999: General Administration and Management. Administration, organization, and management of the Department of the Navy, including general personnel matters (concerning *both* civilian and military personnel), records management programs, security, external and internal relations, audiovisual management, law and legal matters, office services, office automation, and publication and printing matters.
- 6000–6999: Medicine and Dentistry. Medical matters such as physical fitness, general medicine, special or preventive medicine, dentistry, and medical equipment and supplies.
- 7000–7999: Financial Management. Financial administration of the Department of the Navy, including budgeting, disbursing, accounting, auditing, contract auditing, industrial and other special financing matters, and statistical reporting.
- 8000–8999: Ordnance Material. All types of ordnance material and weapons, including ammunition and explosives, guided missiles of all types, nuclear weapons, fire control and optics, combat vehicles, underwater ordnance materials, and miscellaneous ordnance equipment.
- 9000–9999: Ships Design and Material. Such matters as the design and characteristics of ships, and to ships material and equipment.
- 10000–10999: General Material. Those general categories of materials not included in the specialized material groups (such as ordnance and ships). It includes personnel material, general machinery and tools, audiovisual equipment and accessories, and miscellaneous categories including metals, fuels, building materials, electrical and electronic categories, and diving and hyperbaric systems equipment.
- 11000–11999: Facilities and Activities Ashore. Ashore structures and facilities, fleet facilities, transportation facilities, heavy equipment, utilities and services, and other similar subjects.

- 12000–12999: Civilian Personnel. Includes subjects relating solely to the administration of civilian personnel. (Military personnel subjects are included in the 1000 series. General personnel subjects relating to both civilian and military personnel are included in the 5000 series.)
- 13000–13999: Aeronautical and Astronautical Material. Aeronautical and astronautical material, including parts, accessories, and instruments; special devices; armament; aerological equipment, weapons systems, types of aircraft; and astronautic vehicles.

These codes are assigned to various kinds of documents for identification and filing purposes. If you have a working knowledge of them, or if you keep a list of them handy, you can tell something about a document just by the assigned code. For example, if you received a document with the number 12432.1 on it, you would know that it had something to do with civilian personnel because of the 12000-series number assigned. Similarly, the number 4355 would have something to with logistics (because it falls between 4000 and 4999), and 1920 must have something to do with military personnel.

If you look in the SSIC Manual, you will see that the numbers are further broken down to assign more detailed subject matter to the numbers within each series. For example, within the 12000–12999 series (covering Civilian Personnel Records), you would see the following breakdowns:

12000–12099	General Civilian Personnel Records
12100–12199	Office of Personnel Management Records
12200–12299	Personnel Provisions Records
12300–12399	Employment Records
12400–12499	Employee Performance
12500–12599	Position Classification, Pay, and Allowances
12600–12699	Attendance and Leave Records
12700–12799	General Personnel Relations and Services Records
12800–12899	Insurance and Annuities Records
12900–12999	Miscellaneous Records

From this list, you can see that the previously mentioned document with the number 12432.1 on it must have something to do with "Employee Performance" because it falls between 12400 and 12499.

The SSIC Manual further breaks down the numbers as follows:

12430	Performance Management Records
12431	Withholding of Within-Grade Increases Records
12432	Unacceptable Performance Action Records1
12450	General Employee Recognition and Incentives

You can see from the list above that 12430 deals with the subject of "Performance Management" and 12432 breaks the subject down more specifically to deal with "Unacceptable Performance Action." You can also see that there are many numbers (12433–12449) that have no assignment. These numbers may be assigned specific subjects in the future if the need arises. These "missing" numbers can also be used *now* by a command to issue its own directives to cover a specific (related) subject. For example, a command might decide to create a library of leadership books and articles covering various methods of dealing with unacceptable performance. The command might assign the number 12433 to the document establishing this special library, because the subject would seem to fall within this area but there is no number already covering it specifically.

DIRECTIVES

You have probably heard of Navy Regulations. And you might reasonably expect it to be a huge document that spells out the rules and procedures for running the Navy. But this is hardly the case. You could read these regulations in an afternoon and, though they are important, only a rather small number of topics are covered. Among this rather odd assortment of topics are such things as defining the positions of the SECNAV and the CNO (as well as the Commandant of the Marine Corps), establishing precedence among officers, defining standards of conduct, and the proper observance of ceremonies and customs.

So how does one know how to do the thousands of other things that are necessary to running such a large and complex organization? How does one know the proper means of keeping aviation ordnance records, or the rules for conducting proper psychological operations, or the procedures for conferring incentive awards, or the requirements for declassifying secret documents, or the means for effectively using the frequency spectrum, or the correct way of attaching collar devices to a uniform shirt, or. . . . The need for some means of detailing rules, procedures, policies, and the like is obvious. And it does not take a rocket scientist to realize that all of this information needs to be organized in some manner that is comprehensive (translated: *huge*) and yet accessible. That's where the Navy Directives System comes in.

Before delving into this new world of complicated numbers and alien concepts, keep in mind what a gigantic task it is to try to organize so much information; keep in mind that it was conceived long before software was a gleam in Bill Gates's eye; and if you are still not convinced that this system is the best that could be devised, *suggest your own*—and be sure to include an efficient means of transitioning from the existing one to your new one!

If I seem to be apologizing too much in advance, it is because the Navy Directives System is intimidating at first, and without a proper introduction it can be an unfriendly beast. Without keeping in mind what it is trying to accomplish, one is tempted to frequently utter impolite phrases and wonder who the sadist was that dreamed it up. In truth, once you get beyond the alien nature of the beast, you will be able to see a kind of logic in it and perhaps even appreciate the fact that it does manage to organize an incredible amount of information into a navigable maze.

Instructions and Notices

Two kinds of directives are issued by commands within the Navy Directives System. Notices are used to convey temporary (short-lived) information, and instructions are the means of establishing policies and procedures that are of a more permanent (long-term) nature. For example, if the Bureau of Naval Personnel wanted to announce the procedures for a one-time shiphandling contest among junior officers, a notice would be used. But if the bureau wanted to establish an *annual* shiphandling competition, an instruction would be issued.

Notices often contain a self-cancellation date, or their short-lived nature is evident within the text. For example, a notice inviting applicants for a one-time shiphandling competition might have deadline for application included, so it would be obvious that the notice was defunct once that date had passed. Notices usually remain in effect for less than six months and never for more than a year. Instructions, however, remain in effect until superceded by a revised version of the instruction or until formally cancelled by a separate document.

Instructions and notices are issued at virtually all levels of command, from the Secretary of the Navy on down to individual units. The good news is that you do not need to be familiar with those directives issued by commands outside your chain of command; the bad news is that you are responsible for being familiar with those directives issued by your command *and* for all those issued by commands that are above yours in the chain of command. So if you work in the CNO's office, you will be guided by a lot fewer instructions and notices than someone who is assigned to an aircraft carrier in the Atlantic Fleet—that individual must be familiar with the instructions and notices issued by the ship, by the group and fleet commanders, and on up to the CNO and SECNAV (see chapter 2, "Navy Organization").

Identification of Directives

Both instructions and notices are identified by the issuing authority's title, often abbreviated (as in "SECNAV" for Secretary of the Navy, or "OPNAV" for the CNO's office), followed by "INST" if it is an instruction or "NOTE" if it is a notice, and finally an identifying number that can actually tell you something about the subject matter (see the explanation of the SSIC system earlier in this chapter). The date of issuance is also an important identifying component for notices.

An appropriate example is the governing directive for all of this that is issued by the CNO. The title is *Navy Directives Issuance System* and it is identified as:

OPNAVINST 5215.17

It is important to note that there is no other directive with that same identification. There might be a directive issued by another command with the same number, but then it would have a different originating command, such as SECNAVINST 5215.17 or COMPACFLT 5215.17. There might be a notice with the same number (less the decimal as explained below), such as OPNAVNOTE 5216. There might also be another directive with a very similar number, such as OPNAVINST 5215.18, but not another that is identical. That combination of command identifier, type of directive (INST or NOTE), and that exact number is unique to this one directive in all of the U.S. Navy. This is of course a good thing—avoids confusion.

THE DECIMAL PART

By now, you must be wondering why there is a decimal point and the number "17" appended to the SSIC code 5215 in the example above. If we checked the SSIC Manual, we would find the following:

5214	Reports Management
5215	Issuance Systems (Include Directives)
5216	Correspondence Management

We can see that OPNAVINST 5215.17 deals with issuance systems, but we do not see any decimals. What the decimal appendage tells us is that this particular directive is the seventeenth one issued by this command (OPNAV) on that particular subject. You could reasonably expect to find other documents identified as OPNAVINST 5215.16, OPNAVINST 5215.15, and so on, but do not be surprised if you do not; in this ever-changing Navy, directives with some or all of those previous numbers may have been cancelled.

LETTER APPENDAGES

You will often see a letter added to the end of the identifier. This tells us that the directive has been revised. If there is no letter appended, then you know the directive is the original and has not been revised. OPNAVINST 5215.17 has no letter following the number, so it is considered the original version. If the Chief of Naval Operations decides to change some things in this directive but keep it essentially the same, a new directive would be issued with the new identifier OPNAVINST 5215.17A. Another revision would be identified as OPNAVINST 5215.17B and yet another would be OPNAVINST 5215.17C, and so on. If you encountered an instruction with the identifier 6224.1D, you would know that it was the fifth version of that particular instruction (original plus four lettered revisions: A, B, C, and D). Once a revised version is issued, the previous one is superceded and should be discarded.

IDENTIFYING NOTICES

Because they are temporary rather than long-term, notices are identified differently than instructions. They still use an SSIC but you will not find a decimal or an appended letter attached. They are uniquely identified by a date instead. Despite the shared usage of SSICs, notices are easily discernible from instructions by the word "NOTE" following the issuing command's identification.

So if you encountered a directive with the identifier OPNAVNOTE 5215 of 30 August 2006, you would know by the word "NOTE" that it is a notice and not an instruction, and you would therefore know that it is of interest for a relatively short period of time after the date of issue (probably less than six months, as explained earlier). You would also know that it had something to do with the issuing of directives because of the SSIC 5215 (the same as the OPNAVINST 5215.17 discussed earlier).

PUTTING IT ALL TOGETHER

In the section discussing SSICs above, we encountered the instruction "SEC-NAVINST 12432.1" that dealt with the unacceptable performance of civilian employees. We know that this is a directive issued by the Secretary of the Navy ("SECNAV"), that it is a directive of long-term duration ("INST"), that it is the first directive issued on this subject (".1"), and that it has never been modified since it was issued (no letter following the decimal number).

If we encountered a directive identified as SECNAVNOTE 12432 of 26 October 2001, we would know that it was also issued by SECNAV, that it was something of relatively short duration ("NOTE"), that is had something to do with the unacceptable performance of civilian employees (SSIC 12432), and that it is no longer of any interest (more than a year has passed since its issue date). Perhaps it was the announcement of a public flogging date.

A Few More Things to Know About Directives

Official directives are essential to the smooth running of the Navy. They deal with everything from the very important (such as establishing rules of engagement in a combat zone) to the rather trivial (such as how wide the margins must be when creating a directive). Though they are a bit intimidating at first and they are often terribly written (more recent ones are generally less stilted than the older ones), they often contain very useful information.

KEY WORDS

Both instructions and notices are written using specific rules and formats that are spelled out in OPNAVINST 5215.17. Paragraphs will have identifying words at the beginning, like "Purpose" or "Background." This formality is actually useful in that you can generally get right to the meat of the directive by looking for those paragraphs identified with words like "Purpose," "Action," or

"Responsibility" and avoiding those paragraphs beginning with words like "Background" and "Authority."

Even this does not always work. Though the "Purpose" paragraph of SEC-NAVINST 11100.8A is fairly understandable (*"Purpose.* To assign responsibility within the Department of the Navy [DON] for the landlord responsibilities, including the planning and execution of the day-to-day maintenance and long-term care of the Vice President's Residence"), the same paragraph of OPNAVINST 5215.17 is less helpful (*"Purpose.* Per reference (a), enclosure (1) promulgates policies, responsibilities, and standards for the administration of subject system. References (b) and (c) provide additional guidance").

INTERNET EFFICIENCY

In the old days (not so very long ago), directives were typed up, reproduced, and distributed to all who needed them (often amounting to hundreds—even thousands—of pounds of paper). Today, the Navy is saving a great many trees while providing a far more efficient system. Most Navy directives are now maintained on Web sites, which makes them generally more accessible and more likely to be current.

For example, all of the SECNAV directives and those of the OPNAV can be found at the Navy Electronic Directives System Web site (http://doni. daps.dla.mil). You can find a directive by clicking on the button labeled "Navy Directives Index."

At that same Web site, you will find a link to an electronic version of Navy Regulations, as well as links to "New Directives," "Recently Cancelled Directives," and a button called "Subscribe to New Directives," which allows you to obtain e-mail notifications of updates as they occur. Compare that to the old system of having to go to an office where a (hopefully current) library of instructions and notices was kept on shelves in rows of ugly green loose-leaf binders, stuffed with dog-eared pages!

You will also see a link to "SECNAV Manuals." These are instructions that are large and comprehensive enough to qualify as manuals. Examples are the SSIC Manual and the Department of the Navy Forms Management Manual. Note that these instructions are kept separately at this link, rather than with all the other SECNAV instructions. And just in case you were getting the hang of all this standardization, these instructions are exceptions to the usual format in that they are uniquely identified by dropping the "INST" and adding an "M" (for "manual") just before the SSIC, so that the SSIC Manual becomes SECNAV M-5210.2 and the Department of the Navy Forms Management Manual is SECNAV M-5213.1, for example.

USMC DIRECTIVES

In the event that you must deal with directives issued by our sister service, the Marine Corps, it is helpful to know that what the Navy calls "Instructions" (long-term directives), the Marines call "Orders." And the Navy's short-term "Notices" are called "Bulletins" in the Marine Corps.

NAVY CORRESPONDENCE

The Navy applies to its written correspondence the same precision it once applied to the setting of sails to maximize the propulsive power of the wind. These exacting procedures may stifle creativity, but they also ensure a degree of reliability that would not be guaranteed were such stringent procedures not prescribed.

The guiding directive is the Department of the Navy Correspondence Manual (SECNAVINST 5216.5D). Like most government manuals, this one is no thrill to read, but it is an important one with really useful information and guidance that should be on your desk or linked on your computer desktop, particularly if you must write for the Navy. For a more user-friendly source that gives practical advice on naval writing, refer to *The Naval Institute Guide to Naval Writing* by Robert Shenk.

A detailed discussion of naval correspondence is beyond the scope of this book (the Navy Correspondence Manual is more than a hundred pages long), but we will focus on some of the more important aspects in the pages that follow.

Format of Standard Navy Letters

If you are familiar with a basic business letter format, you will find that Navy letters have essentially the same kind of information, but there are some added features and the format is noticeably different. Both kinds of letters have in common a letterhead, the date, the addressee, and the signature, but a standard Navy letter differs in a number of ways.

Figure 4.1 is a sample standard Navy letter. The format you see here is generally the same for all Navy letters (although some may have additional

2310
Ser N3/103
30 August 2007

From: Commanding Officer, Naval Test Wing Atlantic (VX-1)
To: Commander Naval Air Warfare Center Weapons Division, Point Mugu, CA (Code 327100E)

Subj: ORGANIZATIONAL MESSAGE RELEASE AUTHORITY VIA NAMDRP DEFICIENCY REPORT (DR) WEBSITE

Ref: (a) NTP 3(J) Telecommunications Users Manual

Encl: (1) Sample NAMDRP Deficiency Report
(2) Message Form

1. Per ref (a), the following personnel are authorized to release organizational messages for this command using the NAMDRP Deficiency Report (DR) form using enclosures (1) and (2) as guidelines.

Name	Rank/Rate
David Jones	LT
Stephanie Decatur	AT1
W.T. Door	AE1
Corey Spondents	ATCS

2. At least one of the named individuals will be available for releasing messages at all times.

a. Personnel will be given specific rotational assignments.

b. A monthly bill will be maintained by the senior watch officer.

3. This list supercedes all previous authorizations.

C. C. Garrett
Christopher C. Garrett

Figure 4.1 A Typical Naval Letter

features added, such as classification markings if the letter contains information that must be safeguarded from enemy eyes).

Note that in the upper-right hand corner there is a four-digit number ("2310" in figure 4.1). If you are a quick study, you may have correctly guessed that this is an SSIC that can be used to ensure that the letter is filed (and can later be retrieved) by both sender and receiver according to its subject matter.

The serial number below the SSIC is assigned to identify who within the command originated the letter and as an internal means of keeping track of outgoing correspondence. These numbers may vary considerably from command to command and are primarily of interest to clerical personnel.

The date should always be in military format (day-month-year) on a standard Navy letter.

There can be little doubt who is the originator of this letter because there is both a letterhead identifying the command and a "From" line. Standard Navy letters have both "From" and "To" lines ensuring that it is clear who is sending and who is meant to receive the communication. When you think about it, this is more logical than the format of the standard business letter that we all had to get used to when young.

Additional features of a standard Navy letter are as follows:

- "Via" ensures the appropriate levels of the chain of command are kept informed. *Optional*—used only if there are intermediate levels of command involved.
- "Subj" (subject) makes it clear what the letter is about at a quick glance (for those of us who do not have all SSICs committed to memory).
- "Ref" (reference[s]) lists (with alphabetic identifiers in parentheses) any other letters or documents that can or should be referred to in order to better understand the contents of this letter. *Optional*—used when necessary or helpful, but not required if this letter can be understood without additional references.
- "Encl" (enclosure[s]) lists (with numerical identifiers) any documents that are being included with this letter. *Optional*.

Note that the paragraphs in the text are all numbered and are *not* indented. Sub-paragraphs are indented and identified by lowercase letters (a, b, c, etc.). If there are sub-sub-paragraphs, these are further indented and numbered with parentheses around them. The Correspondence Manual prescribes further breakdowns of sub-paragraphs.

This standard format may seem a bit over-prescribed, but once you become comfortable with it, you will realize that it really does help you find what you are looking for quickly. Unlike your credit card statement, which tries to hide much of the most important information, Navy letters are much more forthright and helpful. Just reading the subject line can often tell you whether a letter is of interest to you. Seeing who the letter is from (and to whom it is addressed) can give you some idea of the letter's importance. The cited references can often be useful in further understanding what the letter is about.

If you have to compose a Navy letter, use one of the sources mentioned earlier (the Navy Correspondence Manual or *The Naval Institute Guide to Naval Writing*). Another effective way to do it is to use actual letters as samples. When I was a neophyte Navy letter writer, I kept a file of samples of the different types I encountered and used them as models for composing my own.

Other Forms of Navy Correspondence

There are a number of other forms of Navy correspondence that you may encounter—all of which follow conventions described in the references mentioned above. There are informal memos and more formal things called a "memorandum-for" and a "memorandum for the record." You might encounter a point paper, talking paper, trip report, or a plan of action and milestones (POA&M). If you are familiar with the standard Navy letter, most of these will not seem too alien.

NAVAL MESSAGES

Naval messages are used by virtually every Navy command to send important, official information quickly and reliably. Messages must be officially released by the commanding officer or his or her specifically designated representative(s) before they can be transmitted. Naval communications facilities, whether they are on a small ship or at a large communications station, ensure that officially released messages are accounted for in ways that ensure that all intended recipients get the messages they are supposed to receive.

A naval message can be as simple as a one-line request for an extra tug from a ship to a waterfront port facility or it can be a message released by the Secretary of the Navy or the Chief of Naval Operations that is intended for every person in the Navy. In fact, if you see a message marked "ALNAV" it is indeed from the SECNAV, intended for "all of the Navy (and the Marine Corps)," or if you see one marked "NAVADMIN" it is an administrative message from the CNO and is also intended for broad distribution. These messages are sequentially numbered so that everyone can be sure they did not miss one.

If you encounter a naval message, it is a daunting experience at first, but understanding the format and knowing where to look for specific pieces of information makes the experience tolerable.

Naval messages grew out of the days of Morse code radio transmissions and evolved into the era of radio teletype. When these came into being they were marvelous innovations that allowed the Navy to communicate in ways that were never possible before. A message could be sent to virtually the entire Navy in a very short period of time. A huge communications infrastructure grew out of this capability, involving a worldwide communications net of communications stations, and one unfortunate but inevitable side effect was that as capacity grew, so did the amount of message traffic. Very exacting procedures were put into place to ensure that this increasingly large number of messages got to the intended recipients in a timely manner.

As communications technology continues to develop, this system will probably be phased out and all message traffic will be transmitted via an e-mail type system that will shift message-handling functions away from manpower intensive communications centers to the user's desktop.

Reading Naval Messages

Naval messages can be very short or very, very long, but they all have certain things in common. When reading a naval message, look for clues in finding the parts that are important to you. Figure 4.2 is a real naval message (edited for space considerations) that the Chief of Naval Operations sent out in the aftermath of Hurricane Katrina in the fall of 2005. **The numbers to the left of the message are there for our reference purposes and would not be there in a real naval message.**

At first look, this message appears to be more alphabet soup than an intelligible communication, but looking again with a trained eye will reveal an impressive message that conveys compassion, inspiration, and leadership in a time of crisis.

Note that naval messages are all in uppercase letters; this stems from the days when teletype machines were the primary means of sending naval messages. As

```
1    RAAUZYUW RUEWMFU0966 2622350-UUUU—RUCRNAD.
2    ZNR UUUUU ZUI RUENAAA0966 2622350
3    R 192346Z SEP 05 PSN 675154K28
4    FM CNO WASHINGTON DC
5    TO NAVADMIN
6    ZEN/NAVADMIN @ AL NAVADMIN(UC)
7    INFO ZEN/CNO CNO
8    BT
9    UNCLAS
10   NAVADMIN 236/05
11   MSGID/GENADMIN/CNO WASHINGTON DC/DNS//SEP//
12   SUBJ/TAKING CARE OF OUR OWN//
13   REF/A/GENADMIN/CNO WASHINGTON DC/161133ZSEP2005//
14   AMPN/REF A IS TASK FORCE NAVY FAMILY PLANNING ORDER//
15   RMKS/1. HURRICANE KATRINA DIRECTLY IMPACTED AN ESTIMATED 18,000 NAVY
16   FAMILIES. IN FACT, MANY OF THE SAILORS PROVIDING RELIEF TO LOCAL
17   CITIZENS ARE IN NEED OF RELIEF THEMSELVES. MOST HAVE LOST
18   SOMETHING; SOME HAVE LOST EVERYTHING.
19   2. WE NEED LONG-TERM SOLUTIONS. THAT'S WHY I ORDERED THE ESTABLISHMENT OF TASK
20   FORCE NAVY FAMILY (TFNF). TFNF WILL CONDUCT FULL SPECTRUM COMMUNITY SERVICE
21   OPERATIONS TO PROVIDE A RAPID AND COORDINATED RETURN TO A STABLE ENVIRONMENT FOR
22   OUR AFFECTED NAVY FAMILY. THAT'S THE MISSION.
23   3. AND WHEN I SAY FULL SPECTRUM, I MEAN IT. AS STIPULATED IN REF
24   (A), FULL SPECTRUM COMMUNITY SERVICE OPERATIONS WILL INCLUDE BUT ARE
25   NOT LIMITED TO: 1) FULL ACCOUNTING OF AFFECTED NAVY FAMILY MEMBERS;
26   2) AVAILABILITY OF TEMPORARY HOUSING; 3) WAY AHEAD FOR PERMANENT
27   HOUSING WHERE AUTHORIZED; 4) FINANCIAL ASSISTANCE AND COUNSELING; 5)
28   RETURN TO SCHOOL FOR CHILDREN; 6) TRANSPORTATION OPTIONS FOR
29   RELOCATION, WORK AND SCHOOL; 7) ACCESS TO HEALTH CARE SERVICES; 8)
30   ACCESS TO PASTORAL AND FAMILY COUNSELING SERVICES; 9) ACCESS TO
31   CHILD CARE; 10) ACCESS TO LEGAL SERVICES, INCLUDING CLAIMS SUPPORT;
32   AND 11) EMPLOYMENT SUPPORT.
33   4. JUST TO BE CLEAR, THE NAVY FAMILY CONSISTS OF: NAVY SERVICE
34   MEMBERS (ACTIVE AND RESERVE, OTHER SERVICE MEMBERS ASSIGNED TO NAVY
35   COMMANDS OR TENANTS ON NAVY INSTALLATIONS PENDING CONCURRENCE OF
36   THEIR RESPECTIVE SERVICES) AND THEIR FAMILIES; NAVY RETIREES AND THEIR FAMILIES;
37   CIVILIAN EMPLOYEES OF DEPARTMENT OF THE NAVY AND THEIR FAMILIES; AND MAY INCLUDE
38   CERTAIN EXTENDED FAMILY MEMBERS (DEFINED AS PARENTS, PARENTS-IN-LAW, GUARDIANS,
39   BROTHERS, SISTERS, BROTHERS-IN-LAW, SISTERS-IN-LAW) OF DECEASED, INJURED,
40   OR MISSING NAVY SERVICE MEMBERS, NAVY RETIREES OR DON CIVILIANS WITHIN THE JOA;
41   FAMILY MEMBERS IN THE JOA OF NAVY SERVICE MEMBERS/CIVILIANS.
42   5. I WANT THE NET CAST WIDE, AND I WANT IT HAULED IN OFTEN. THERE
43   ARE PEOPLE HURTING OUT THERE — OUR PEOPLE AND THEIR LOVED ONES —
44   AND WE WILL DO ALL WE CAN TO ALLEVIATE THEIR PAIN. I LIKEN IT TO A
45   MAN OVERBOARD. YOU SHIFT THE RUDDER OVER, GO TO FLANK SPEED, AND
46   PLUCK THE SAILOR OUT OF THE WATER. IN MY VIEW, WE VE GOT NEARLY
47   45,000 PEOPLE IN THE WATER RIGHT NOW, AND WE RE GOING TO PICK THEM
48   UP.
49   6. HURRICANE KATRINA DEVASTATED CITIES AND TOWNS. IT TOOK LIVES.
50   BY DAMAGING OUR BASES IN THAT REGION, IT EVEN CHIPPED AWAY AT SOME
51   OF OUR COMBAT CAPABILITY. BUT IT DID NOT DESTROY THE HUMAN SPIRIT.
52   IT DID NOT DESTROY THE NAVY FAMILY. NO STORM CAN WIPE THAT OUT. WE
53   WILL STAND BY THE NAVY FAMILY AS THE NAVY FAMILY HAS STOOD BY US. I
54   KNOW I CAN RELY ON YOUR SUPPORT.
55   7. ADM MIKE MULLEN SENDS.//
56   BT
57   #0966
```

Figure 4.2 A Typical Naval Message

you wade into the message, keep in mind that some of it is there primarily for the communicators who must create, transmit, and account for the message. If you see a line preceded by a three-letter code (such as "ZNR" and "ZEN"), the information on that line is for the communicators and you can ignore it.

Key Elements

The main things to look for in a naval message as you scan from top to bottom are the following:

- FM
- TO
- SUBJ
- Single sequential numbers (marking text paragraphs)

FM

The line that begins with "FM" (line 4 in our example in figure 4.2) is short for "from" and tells you who sent the message; in this case, the CNO.

TO

Likewise, the "TO" line (line 5 in our example) tells you whom the message is meant for.1 It might designate one or more individual addressees (like USS PREBLE or DESRON 15) or it might use a more efficient collective address as in this case ("NAVADMIN" means it is a special message from the CNO that is meant for the entire Navy). If there is more than one addressee, each is on a separate line, and sometimes the list can be quite long if the message is meant to go to a lot of different units.*

SUBJ

Although the "SUBJ" line (line 12 in figure 4.2) is not at the top of a naval message, it can be a real time saver if it is done right because it reveals the subject of the message. Train your eyes to look for that first, and many times you can tell right off whether the message is of interest to you. If you are an engineer working on submarine sonar systems and the "SUBJ" line reads "AVIATION

---

*Sometimes messages will have another line that follows the TO line(s), and it is identified by "INFO," which stands for "information addressee(s)." If a message is addressed "TO" someone, they are considered "action" addressees and they are expected to do something as a result. "INFO" addressees are not expected to take any action; they are being kept informed of the message's contents.

FUEL INVENTORIES," chances are you can ignore it. Be careful, however. Subject lines are not always clear and may not accurately reflect the message content (or all of it); naval messages are drafted by human beings, after all. A line with a SUBJ of "PREPARING FLAG BRIEFS" may not be about laundry.

TEXT

To read the important part of a naval message—to get to the meat of the matter—look for the text, which is easily identifiable because *the paragraphs are numbered* with a period following. You can see in our example that a little way in on line 15 is a "1." That is the first paragraph of the message. The second paragraph is identified by a "2" (line 19 in figure 4.2), the third by a "3" (line 23) and the others at lines 33, 42, 49, and 55. Be careful not to confuse the numbers with a single parenthesis following (beginning at line 25) with paragraph numbers. These are numbered items on a list within paragraph 3.

The Date-Time Group

Near the top of every message (line 3 in figure 4.2) is a thing called a "Date-Time Group" (DTG), which tells you the approximate time and date that the message was sent. It is important because it uniquely identifies the message and because you can tell when the message was sent (something that will sometimes help you understand the relevance of a message).

The DTG always appears in the same format: six numbers followed by a "Z" and then a three-letter abbreviation for the month and a two-digit number for the year. In most cases all you will care about is the date that the message was sent; the month and year are obvious but the day is less so, because it is represented by the first two digits of the six that appear before the "Z."* It is always in two digits, so the seventh of the month would appear as "07." So, looking at the DTG on our example message, we know that it was sent on 19 September 2005.

One more thing to possibly keep in mind about the DTG is that each message sent by a particular originator has a unique DTG and this is used to reference that particular message. In other words, no other message originated by the CNO will have this same DTG, and if he or she or anyone else wants to refer to this message in another message, they will do so by the DTG.

---

*The next four numbers are the time (always in twenty-four-hour time) and the "Z" indicates that the time is not local but is UTC (Coordinated Universal Time—which used to be called Greenwich Mean Time). This is usually important only to professional communicators.

Message Classification

Depending upon where you work and what you do, you may have access to classified material. In the next section of this book we will discuss this in more detail, but in terms of naval messages, you can tell what the classification is (or is not) by looking between the addresses and the subject line. Line 9 in figure 4.2 is the message classification; in this case, the message is unclassified; you can tell by the abbreviation "UNCLAS." If the message were classified it would have the words "CONFIDENTIAL," "SECRET," or "TOP SECRET" instead, and you would have to have the appropriate security clearance to read it.

INFORMATION SECURITY

It is obvious that some information should be protected from our enemies or our potential enemies. Such things as the contents of our contingency war plans or the technological details of a unique weapons system are examples of the kinds of information that can be harmful to the nation if it falls into the wrong hands. Some kinds of information are more potentially harmful than others, so it makes sense to have levels of classification with each higher level requiring more protection.

Information security can be a complex subject, involving how things are classified, how they are physically protected, how they are accounted for, how they are declassified, and so on. Details are provided in a directive issued by the Secretary of the Navy called the Department of the Navy Information Security Program Regulation (SECNAVINST 5510.36). Fortunately, the basics of information security are rather simple.

Security Classifications

The level of classification determines how much protection a piece of information requires. There are four levels of classification, each of which indicates the anticipated degree of damage to national security that could result from unauthorized disclosure:

Top secret	Exceptionally grave damage
Secret	Serious damage
Confidential	Damage
Unclassified	No damage

All classified material—such as publications, software, equipment, or films—must be plainly marked or stamped with the appropriate classification designation. Following the classification, some material may have additional markings that signal extra precautions in handling. For example, the marking "restricted data" means that the material pertains to nuclear weapons or power.

There is another category of government information, "for official use only" (FOUO). This is not classified information because it does not involve national security, but it is information that could be damaging in other ways and cannot, therefore, be divulged to everyone. Results of investigations, examination questions, bids on contracts, and so on, are "privileged information" and are kept from general knowledge under the designation FOUO.

Security Clearance

Before a person is allowed to have access to classified information, he or she must have a security clearance. If, in doing your job, you need to work with information that is classified "secret," you must first obtain a secret clearance. If all you will need to see is confidential information, you will be assigned a confidential clearance. In other words you will be given the highest level of clearance that you will need but no higher. So you should not take it personally if your clearance is lower than the person on the next deck up; clearances are not merely a measure of your trustworthiness, they are also tied to what you *need* in order to do your job. If you are ever denied the access you need, that is a cause for concern, but as long you receive the level of clearance required for the performance of your assigned duties, you should be satisfied.

Investigations

Before you can be granted a security clearance, an investigation is conducted into your background to make certain that you can be trusted with classified information. Government agents will look into your past records and question people who have known you. This process takes a while, so you may be given an "interim" clearance based upon some preliminary investigating before your "final" clearance comes through. The word "final" in this case means that the routine investigating is over and that you have been granted the clearance you need. It is not "final" in the sense that it cannot be taken away. In order to get a security clearance, you must be trustworthy, of reliable character, and able to show discretion and good judgment.

In order to receive a clearance, the investigation must conclude that you are a person who is not only loyal to your country but that you are also able to meet the standards for a position of trust and confidence. Conduct such as drug abuse, excessive drinking, or financial irresponsibility can lead to denial of clearance. A clearance may also be denied or revoked because of mental or emotional conditions, general disciplinary causes, falsification of official documents, or disregard for public laws or Navy regulations. Should you involve yourself in any of the disqualifying activities mentioned above (such as drug use or financial irresponsibility) your clearance may be revoked. Also, if you change jobs, or for some other reason no longer need to have access to the same level of classification, your clearance will be reduced or removed without prejudicing your future eligibility. Commanding officers may reinstate your previous clearance as the need arises.

Access and Need to Know

As stated above, security clearances grant you access to the level of classification that is necessary for you to perform your official duties. But it is important that you understand the concept of "need to know." Just because you have a secret clearance, that does not give you access to *all* secret material. Your secret clearance allows you to see all the secret material you need to know in order to do your job, but it does *not* entitle you to see information classified at that level that has nothing to do with your job.

Safeguarding Classified Information

Classified information or material is discussed, used, or stored only where adequate security measures are in effect. When removed from storage for use, it must be kept under the continuous observation of a cleared person. It must never be left unattended.

Security Areas

Spaces where classified materials are used or stowed or that serve as buffers are known as security areas. Some areas are more sensitive or are more likely to risk compromise than others, so to meet these varying needs, a system has been developed to identify security areas properly. The government has established three types of security areas and identified them by levels. All three of these areas are clearly marked by signs with the words "Restricted Area." The specific level of the area is not identified on the signs, however.

LEVEL I

No classified material is actually used or kept in a level I space; it is used as a buffer or control point to prevent access to a higher-level security area. A security clearance is not required for access to a level I area, but an identification system is usually in place to control access to the area.

LEVEL II

Classified material is stowed or used in these areas. Uncontrolled access to a level II area could potentially result in the compromise of classified information. Therefore, anyone not holding the proper clearance must be escorted while visiting a level II area.

LEVEL III

Classified material is used in a level III area in such a manner as mere entry into the area risks compromise. An example would be a command and control center where large decision-making displays have classified information posted on them. Only people with the proper clearance and the need to know are permitted access to a level III area. All entrances must be guarded or properly secured.

Disclosure

You are responsible for protecting any classified information you know or control. Before giving another person access to that information, it is your responsibility to determine that the person has the proper clearance *and* a need to know. If you are uncertain whether someone has the proper clearance and a need to know, find out *before* you allow them access. Never tell someone something classified just because they are curious, even if they have the proper clearance. Remember, there are two requirements for someone to have access to classified material: they must have the proper clearance and they must have an official need to know the information.

Voice Communications

Some telephones and radio circuits in the Navy are what we call "secure." This means they are protected by special equipment that encrypts (scrambles) your voice so that an enemy cannot listen in and understand what you are saying. Never discuss classified information over a telephone or a radio circuit unless you know it is secure.

Compromise

Unauthorized disclosure, or "compromise," means that classified information has been exposed to a person not authorized to see it. If you have any suspicion that classified material has been compromised, you must report it immediately.

Material that has been misplaced or lost is considered compromised and must be reported immediately.

Discovery of Classified Information

If you accidentally come across some classified material that has been left unguarded, misplaced, or not secured, do not read or examine it or try to decide what to do with it. Report the discovery immediately and stand by to keep unauthorized personnel away until an appropriate official takes custody of the material.

Stowage and Transport

Classified material may not be removed from the command without permission. Authorized protective measures must be used when classified material is sent or carried from one place to another, and it must be stowed (stored) properly.

Do not, for example, take a classified manual home to study at night. It is admirable that you want to improve your knowledge so that you can do your job better, but you probably do not have the means to transport the material safely, you almost definitely do not have the means to stow it safely in your home, and because you will not have the permission of your commanding officer, you will be in very serious trouble should anything happen to the material.

Threats to Security

Foreign nations may be interested in classified information on new developments, weapons, techniques, and materials, as well as movements and the operating capabilities of ships and aircraft.

The people who collect such information cannot be stereotyped or categorized, which is why they succeed in their work. A foreign intelligence agent collects many odd little bits of information, some of which might not even make sense to the agent, but when they are all put together in the agent's own country they may tell experts enough to be damaging to our national security.

Enemy agents infiltrate social gatherings where they gather important pieces of information merely by listening to the conversations around them or by actively engaging in dialogue with potential targets. Some agents even move into communities where people with classified access live so they can collect information from their neighbors.

Espionage agents prey upon the vulnerabilities of their intended targets. For example, people with relatives in foreign countries can sometimes be intimidated into cooperation with threats to their relatives. People with financial problems or drug habits can be coerced into doing favors for the enemy. Some people may have a desire to feel more important or feel a need for attention and these traits can be exploited by skilled agents.

One devious but effective scenario occurs when a relative stranger offers to solve a problem you are having; if you accept his or her help, and your new friend turns out to be from an unfriendly foreign nation, you are now in an awkward position. Even though you may not have divulged any classified information, your acceptance of favors from someone who suddenly divulges his or her true identity appears suspicious and embarrassing. People who are weaker in character will be tempted to go further and reveal classified information rather than reveal their connection to a foreign agent. This may sound like a scene from a spy movie but, unfortunately, it happens all too often in real life.

Listed below are some ways to prevent being exploited by a foreign agent:

- Don't talk about a sensitive job to people who don't need to know—not even to your family or friends.
- Be careful what you say in social situations. Even seemingly trivial information can be valuable in the wrong hands.
- Know how to handle classified material properly.
- If you have personal problems you feel might be exploited, use the chain of command to solve them. If one of them can't help, go to the chaplain. Chaplains are in the service for a lot more than conducting religious services; they are there to help, whatever your problem is.

Reporting Threats

Report any suspicious contact. If someone seems more curious about your job than seems normal and presses you for information in any way, report it to your superiors. If the person is innocent, no harm will come of it. If the person is guilty, you will have done a great service to your country by calling attention to the incident.

If you are contacted by someone who you are certain is attempting espionage, do not try to be a hero by taking action yourself. Report it!

Report any contact with someone you know who is from a nation that is hostile or potentially hostile to the United States, even if the contact seems innocent. Remember that spies rarely start out trying to get classified information from their targets. If you are unsure whether the nation is considered a potential threat, report it. In matters of security, it is always better to be overly cautious than not cautious enough.

If you feel you can't approach the people in your chain of command, go to the Naval Criminal Investigative Service (NCIS) office. If you can't find one, look in the white pages of the phone book under U.S. Government, Naval Activities.

If you are going to make a report, make a note of the date, time, place, and nature of the encounter. Describe how you were approached and mention who else in the Navy was also approached. Provide names if you know them.

Most people are not naturally suspicious and no one enjoys reporting the activities of others. But if you have been entrusted with information that can help enemies do us harm, you have an added responsibility to safeguard that information to the best of your ability.

QUICKREFS

- Important basic references:*
  U.S. Navy Regulations (NAVREGS): http://neds.daps.dla.mil/regs.htm
- Standard Organization and Regulations of the U.S. Navy:
  http://neds.daps.dla.mil/312032.htm
- Standard Subject Identification Code (SSIC) Manual:
  http://neds.daps.dla.mil/Directives/5210_2.pdf
- Standard Navy Distribution List (SNDL):
  http://neds.daps.dla.mil/sndl.htm

---

*(Note that URLs are provided for most, but these may change. If the URLs provided do not work, use a search engine (such as Google) to find these publications online. Many are available at the DONI Web site (http://doni.daps.dla.mil/) and at the Naval Personnel Command Reference Library site (http://www.npc.navy.mil/ Reference Library/).

- Naval Military Personnel Manual (MILPERSMAN):
  http://buperscd.technology.navy.mil/bup_updt/508/milpers/index_
  milpersman.htm
- Manual for Courts-Martial, United States (MCM):
  http://www.jag.navy.mil/documents/mcm2000.pdf
- Manual of the Judge Advocate General (JAGMAN):
  http://www.jag.navy.mil/documents/JAGMAN2004.pdf
- Joint Federal Travel Regulations (JFTR):
  https://secureapp2.hqda.pentagon.mil/perdiem/jftr(ch1-ch10).pdf (volume
  1) and https://secureapp2.hqda.pentagon.mil/perdiem/jtr(ch1–16).pdf
  (volume 2)
- U.S. Navy Travel Instructions:
  (not available at the time of this writing)
- Department of Defense Military Pay and Allowance Entitlements Manual
  (DODPM):
  (not available at the time of this writing)
- Navy Pay and Personnel Procedures Manual (PAYPERSMAN):
  (not available at the time of this writing)
- Enlisted Transfer Manual (TRANSMAN):
  http://www.sdmcp.org/Regs/EnlistedTransferManual.pdf
- Navy and Marine Corps Awards Manual:
  http://neds.daps.dla.mil/Directives/1650/1650.htm
- Manual of Advancement:
  http://buperscd.technology.navy.mil/bup_updt/508/INSTRUCTIONS/
  143016/1430.16.htm
- The Navy's *SSIC system* provides four- and five-digit codes that match up to
  specific subject areas for identification and filing purposes. These codes can
  be broken down further, but the major breakdowns are as follows:

1000–1999	Military Personnel
2000–2999	Telecommunications
3000–3999	Operations and Readiness
4000–4999	Logistics
5000–5999	General Administration and Management
6000–6999	Medicine and Dentistry
7000–7999	Financial Management
8000–8999	Ordnance Material

9000–9999	Ships Design and Material
10000–10999	General Material
11000–11999	Facilities and Activities Ashore
12000–12999	Civilian Personnel
13000–13999	Aeronautical and Astronautical Material

- *Navy directives* are used to supplement Navy Regulations, providing policies, procedures, and detailed instructions for the efficient running of the service.
- Each command may issue its own directives. You should be aware of those of your command and those from commands senior to yours in the chain of command.
- Permanent directives are called "instructions" and temporary directives are called "notices." Instructions remain in effect until cancelled or superceded by a revised version; notices often have a self-canceling date or their short-lived nature is evident within the text.
- Directives are categorized using SSICs, and they are uniquely identified using modified versions of SSICs (for instructions) and dates (for notices).
- Paragraphs are labeled so that you can quickly locate the most relevant parts: look for paragraphs labeled "Purpose" or "Action."
- NEDS conveniently provides all of the SECNAV and OPNAV instructions and notices at a special Web site that is accessible via the Internet: http://neds.daps.dla.mil/.
- USMC directives serve the same purposes but are called by different names: instructions are "orders" and notices are "bulletins" in the Marine Corps.
- The *standard Navy letter format* contains the following elements: *From*, *To*, Via, *Subj*, Ref, Encl, *Text*, *Signature* (elements in italic are always present; others frequently appear).
- The key elements to look for in a *naval message* are
  —*FM* (which tells you who is sending the message),
  —*TO* (which tells you who the intended recipients are),
  —*SUBJ* (which tells you what the message is about), and
  —*Single sequential numbers* (which mark the paragraphs in the text).
  —*Security classifications* for the safeguarding of sensitive information are based upon the degree of damage that can be caused to the nation if an enemy had access to it:

Top secret	Exceptionally grave damage
Secret	Serious damage
Confidential	Damage
Unclassified	No damage

- To have access to classified information, you must be granted a *formal clearance* for the level of classification involved and have a *need to know* that particular piece of information.
- Certain standards must always be applied when dealing with classified information:

  —Never discuss it over a telephone line or radio circuit that is not secure.

  —Make sure it is properly stored when not in use.

  —Transport it only for official reasons and keep it safeguarded at all times.

  —Report lost or found classified material immediately.

  —If some unauthorized person has been exposed to classified material, report it immediately.

  —If unauthorized personnel make any attempts to obtain classified information from you, report it immediately.

CHAPTER FIVE

Navy Customs and Traditions

To people on the outside looking in, the Navy's customs and traditions are often fascinating, quaint, endearing, inspiring, and sometimes just plain funny. But for those who are inside the Navy, they can be quite frustrating if they are not understood. One can feel excluded or confused or both when surrounded by strange goings-on that have not been properly explained. In some cases, not understanding the Navy's customs and traditions can lead to needless embarrassment.

Many of these practices (such as saluting) you will merely observe, but some you may want to participate in (such as those customs observed during the national anthem).

In other chapters we will discuss some of the oddities within the Navy (such as the sanctity of the quarterdeck, the use of nautical terminology even when ashore, and the use of bells to tell time). Though it is beyond the scope of this book to go into every custom and tradition in the Navy, in this chapter we will go deeper into the mysterious world of some of the Navy's strange habits and practices, so that you can better understand, and, in many cases, appreciate what is behind them.

THE SALUTE

Perhaps the most curious of all customs in the Navy—and the other services as well—is the salute. What makes it even stranger is that all the services do not do it exactly the same!

It is a centuries-old custom and, although the origin of saluting is unclear, it probably originated when men in armor raised their helmet visors so they could be identified. The tradition continues today as a means of showing respect among warriors.

In the Navy, salutes are customarily given with the right hand, but there are exceptions. If a person's right arm is injured, or if he or she is using the right hand for some other military purpose (such as holding a boatswain's pipe while blowing it), then it is considered appropriate to salute with the left hand. However, people in the Army and Air Force never salute left-handed.

To make things more confusing, a Soldier or Airman may salute uncovered (without cap on), whereas Sailors must be covered if they are going to salute. Marines are more closely aligned with Navy saluting practices than Army or Air Force.

Be aware that these differences in custom among the services may be modified if the circumstances warrant. Consider, for example, if you are in a joint office (where there is a rainbow of uniforms running around), the customs of the predominant service may be observed by all. The old (and *customary*) saying, "When in Rome, do as the Romans do," is good advice.

In most cases, a salute is accompanied by a verbal greeting. For example, when a petty officer meets a lieutenant as they are coming to work, he will salute her and say, "Good morning, Lieutenant Jones." If he does not know the officer's name, "Good morning, ma'am," is appropriate.

Who Salutes Whom

The most general rule for saluting is that *juniors salute seniors*. But there are a few caveats to this.

To begin with, when we say that juniors salute seniors, we really mean that juniors *initiate* the saluting process (in other words, they go first). Salutes between individuals should nearly always be returned, however.

Enlisted personnel will salute officers (because the former are junior to the latter). (See chapter 1, "Military Titles," for an explanation of enlisted and officers.) Junior officers will salute more senior officers. Officers of equal rank should salute one another—truthfully (and sadly) you will often see this custom ignored. In case you were getting lulled into a sense of, "This is too easy," you should know that enlisted people normally do not salute one another, no matter what their relative ranks. And there is even an exception to this last rule: in some cases, enlisted personnel standing watches or carrying out other military duties may salute one another as part of their duties.

These rules span the various services. That is, you will see an Army sergeant salute a Navy commander, or a Navy ensign salute a Marine Corps colonel.

These rules also span international boundaries. You may well see a U.S. Navy petty officer salute a captain in the Chilean navy, or a British private salute a U.S. Coast Guard lieutenant.

Officers in the U.S. Merchant Marine and Public Health Service wear uniforms that closely resemble Navy uniforms, and they too are included in the saluting process.

As mentioned above, sometimes a person's official duty will trump the rules above. A chief petty officer standing an OOD (Officer of the Deck) watch on a ship moored to a pier will be saluted by all personnel arriving on board or departing the ship, no matter what their rank or rate. This is because the chief's temporary status as OOD supercedes his being a chief. There are other exceptions similar to this but a detailed discussion is beyond the scope of this book.

Sometimes salutes are exchanged beyond the rules. *The Bluejacket's Manual* advises Sailors: "Because you are in uniform, young children will often salute you; they rightfully associate saluting with military behavior. Return the salute. The smile you will get in return will make your whole day."

Saluting While Armed

When Sailors have a sidearm (pistol) in a holster, they salute the same as if they were unarmed; the same is true if they are carrying a rifle at "sling arms" (carried on a strap over one shoulder). But when carrying a rifle (other than at "sling arms"), there are three different ways in which to salute.

- *Present Arms* is a salute in itself and is the one most often used. (The rifle is held vertically in front, parallel to the body, trigger guard facing away from the body, with the muzzle about three inches higher than the eyes.)
- When a rifle is at *Order Arms* (butt of the rifle resting on the deck next to the right foot, held in place by the right hand), the left hand can be brought across the body (smartly) and held in a flat saluting posture, touching the rifle just below the muzzle.
- When a rifle is carried at *Right Shoulder Arms* (resting on the right shoulder at about a 45-degree angle, forearm parallel to the deck, trigger guard against the body), it is appropriate to salute by bringing the left arm smartly across the body, keeping the hand flat in appropriate salute fashion, and touching the fingertips to the rifle while keeping the arm parallel to the deck.

When Salutes Are *Not* Exchanged

There are times when other considerations override the desirability of a military salute. Salutes are dispensed with in the following circumstances:

- When engaged in work and saluting would interfere with the work.
- When engaged in athletics or some other recreational activity.
- When carrying something with both hands and saluting would require putting all or part of the load down. A verbal greeting is still appropriate in this case.
- In public places where saluting is obviously inappropriate (such as on a bus or while standing in line at a theater). A verbal greeting is usually still appropriate, however.
- When in combat or simulated combat conditions.
- While at mess (while eating).
- When guarding prisoners.
- While in a military formation. The person in charge will salute for the entire group or, in some cases, will give the order for the entire formation to salute in unison (as during a formal inspection).
- After the first salute of the day has been rendered. In other words, once an individual has saluted someone, that individual is not expected to salute that same person every time they encounter one another. Once per day is all that is expected, except on very formal occasions.

Gun Salutes

In the old days it took as long as twenty minutes to load and fire a gun, so when a ship fired her guns in salute, thereby rendering herself temporarily powerless, it was a friendly gesture. That practice has come down through the years to be a form of honoring an individual or a nation.

The gun salutes prescribed by Navy regulations are fired only by ships and stations designated by the Secretary of the Navy. Salutes are fired at intervals of five seconds, and always in odd numbers. A salute of twenty-one guns is fired on Washington's Birthday, Memorial Day, and Independence Day, and to honor the President of the United States as well as the heads of foreign states. Other high-ranking government officials are honored by a lesser number of guns; for example, the Secretary of the Navy receives a nineteen-gun salute.

Senior naval officers are also honored by gun salutes, and the number of shots fired depends upon their rank. Salutes for naval officers are as follows:

Admiral	17 guns
Vice admiral	15 guns
Rear admiral (upper half)	13 guns
Rear admiral (lower half)	11 guns

Officers below the rank of rear admiral (lower half) do not rate a gun salute. Details of who rates what can be found in Navy Regulations.

FLAGS AND FLAG ETIQUETTE

Flags have always been a part of the Navy. National flags are displayed by ships at sea as a means of identification; signal flags were the main means of communication among ships before the invention of radio; admirals are frequently referred to as "flag officers" because they have their own flags as a symbol of their authority and responsibility; and each of the armed services has its own flag. As with most things in the Navy, flags are used in prescribed and customary ways.

The National Ensign

Military people do not just salute one another. They also salute the American flag. In the Navy, you will hear the American flag more often called "the national ensign" or simply "the colors." A national ensign is technically a special flag that is flown by ships at sea, and many nations have an ensign that is different in some way(s) from the national flag. For the United States, the national ensign and the national flag are the same, but because we are Navy, it has become the custom to use the term "ensign."

In truth, the national ensign is merely a piece of colored cloth. But it is a *symbol* of the freedoms we Americans enjoy and treasure. And it takes on extra significance for Sailors because part of the military job description is a willingness to make great sacrifices; it is the "mere piece of cloth" that is placed on the coffins of fallen warriors.

The National Anthem

Many customs and ceremonies are associated with the national ensign. One that you will have some familiarity with is showing respect during the playing of the national anthem. Just as most Americans stand during the playing of "The Star-Spangled Banner" before a ball game, all naval personnel show similar respect whenever the national anthem is played. This is accomplished by standing at attention and facing the national ensign if it can be seen, or facing in the direction of the music if the ensign is not in sight.

If in uniform and covered (wearing a cap), naval personnel salute from the sounding of the first note to the last. If they are in uniform but uncovered, or if they are in civilian clothes or athletic gear, they will stand at attention during the playing of the anthem but not salute. Men in civilian clothes but wearing caps will remove them and hold them over their hearts. All naval personnel not in uniform and not covered will place their right hands—palm open and against the chest—over their hearts during the playing of the anthem. It is customary for civilians to do this as well.

On military bases, all drivers (military and civilian) of motor vehicles should pull over and stop if traffic safety permits. Driver and passengers sit quietly at attention but do not salute.

Be aware that on some occasions personnel from allied services may be present, in which case a foreign national anthem may follow the U.S. one. The same marks of respect prescribed during the playing of the U.S. national anthem are shown during the playing of a foreign national anthem.

Passing of the Colors

Whenever the national ensign is being carried by a color guard and passes by you, as in a parade, for example, you should stand (if you were sitting). The same rules apply as during the playing of the national anthem: salutes from military personnel in uniform, hat or hand over heart for all (military and civilian) who are in civilian clothes.

Morning and Evening Colors

The ceremonies of hoisting the national ensign (raising the flag) at 0800 (8 am) and lowering it at sunset are called "morning colors" and "evening colors," respectively. These ceremonies take place every day on every Navy shore station in the world. Ships at sea do not observe either of these formal ceremonies, but ships in port—whether moored to a pier or anchored offshore—*do* observe both morning and evening colors.

JACKS

Aboard ships in port, the national ensign is raised on the "flagstaff," a relatively short spar (pole) mounted at the very aft (rear) end of the ship. There is an added factor aboard ship that you will not see on shore: another flag is hoisted simultaneously onto another spar on the bow (located as far forward as possible). This spar is called the "jackstaff" because for most of the Navy's existence, the "union jack" was the flag hoisted here.*

Twice in the Navy's history, the union jack has been replaced by a different flag that has been known by several names: "First Navy Jack," "Navy Jack," the "Dont Tread on Me" flag,** and sometimes "the rattlesnake flag." This flag has thirteen red and white stripes just as on the national ensign, but instead of containing a blue star field, it has a rattlesnake diagonally superimposed and the words "Dont Tread on Me" horizontally across the bottom. This flag first appeared in the fall of 1775, when Commodore Esek Hopkins signaled his fleet to engage the British enemy and directed that they display the "strip'd Jack and Ensign at their proper places."

The first substitution of this flag for the traditional union jack occurred during the nation's bicentennial, when it was flown for the entire year in 1976. The second time began on 31 May 2002 and continues today. At that time, the SECNAV decreed that "during the Global War on Terrorism" the First Navy Jack will be "displayed on board all U.S. Navy ships in lieu of the union jack."

MORNING COLORS

Aboard ship (and on many naval bases), "First call to colors" is sounded on the general announcing system (1MC) precisely at 0755 (7:55 am). When feasible, this is a special bugle call. Few Navy ships today have a bugler aboard as part of the ship's company (crew), as was common in earlier days, so the tradition is often kept alive by using a recorded bugle call. An alternative is for the words "first call to colors" to be passed instead. At the same time, a special yellow and green pennant called the PREP (for "preparatory") pennant is hoisted to the yardarm on ships (or on a yardarm on the main flag pole on a

---

*The union jack is a flag that is the blue star-studded rectangle on the American flag; in other words, if you cut away the stripes from an American flag, you would be left with a union jack. It is also flown from a yardarm when a court-martial is in progress, and it is also used as the President's and SECNAV's personal flag.

**You might have noted that I spelled "Dont" without the apostrophe. This is because that is the way it appeared on the original flag and on replicas today. I don't know if it was spelled that way because of different punctuation rules in the eighteenth century or because the patriots of the day had slept through English grammar classes in school.

base). You will be able to tell youth from experience at this point because the veterans of battles with enemies or the elements will often come out to take part in the ceremony, whereas the young and inexperienced will hurry inside to avoid participating because they have not yet come to appreciate what it symbolizes.

At 0800 (8 am), the bugle sounds "attention" (or a whistle is blown) and the national ensign and jack are hoisted. At that moment, the PREP pennant will be hauled to the dip (lowered to the halfway point) and remain there until the ceremony is completed. While the colors are being briskly hoisted, one of several things will happen:

- A band plays the national anthem (if the ship or shore station has a band).
- A bugler plays "To the Colors" (if the ship or station has a bugler assigned).
- A recording of the national anthem is played over the announcing system.
- A recording of "To the Colors" is played over the announcing system.
- Silence is observed while the colors are being hoisted (if none of the choices above are available).

During colors everyone within sight or hearing renders honors. Navy people outside stop working (or whatever else they were doing when attention was sounded), face the colors, and salute until they hear the national anthem or "To the Colors" end. Although no longer saluting, they will remain at attention until "carry on" (either by bugle, whistle, or voice) is sounded. If they cannot see the national ensign, they will face in the direction the music or whistles are coming from.

Though you would not salute, as a part of the Total Force triad it is appropriate for you to also show respect for the ceremony by standing at attention and rendering honors as described above for the national anthem (removal of hats, etc.).

EVENING COLORS

Sunset is the time for evening colors in the Navy. The exact time of sunset changes depending upon your latitude and the time of year. The ceremony is similar to morning colors. Five minutes before sunset, "First call to colors" is sounded just as in the morning, and the PREP pennant will again be raised to the yardarm. At sunset, the colors ceremony begins when "attention" is sounded on a bugle (in most cases a recording) or when a whistle is blown. PREP is hauled to the dip just as in the morning and the procedures for standing at attention and saluting are the same as in the morning. While the national ensign is being lowered, the bugler (or recording) will play "Retreat" (instead of "To the Colors," as is played in the morning). The national anthem is not normally played at this time. Another difference in the two ceremonies is that at morning colors the national ensign is hauled up smartly (quickly), whereas at evening colors it is hauled down slowly and ceremoniously. "Carry on" will signal the end of the ceremony, just as in the morning.

THOUGHTS

During the colors ceremony, you may of course think about whatever you like, but we who serve this nation as part of the Navy sometimes use this isolated moment to think about what this Navy and this nation are all about, to think about what it is that makes the United States the greatest nation on earth, never perfect but continually striving to be. Whether your contribution is hoisting bombs to the belly of an aircraft or putting words on paper, taking a submarine into the abyss or helping to design it, maneuvering a destroyer or repairing it, it is no exaggeration to say that we all are part of what keeps this nation safe. Though no nation is served well by jingoists or blind patriots, it is not a bad thing to feel a surge of pride at the sight of our nation's flag bursting forth on a morning breeze in all its colorful glory.

And just as morning colors will often cause a great surge of pride as the national ensign makes its dramatic appearance, so evening colors is a time for quiet reflection. As you stand respectfully in the evening twilight, watching the national ensign slowly descend the mast to the haunting notes of a bugle playing "Retreat," do not be surprised if you feel a special bond with your nation and an appreciation for the sacrifices that have been made by people just like you in its defense. It is one of those moments that people in other walks of life never exactly share, and one that you may well remember for the rest of your life, whether you leave the Navy after a short time or remain in service many decades.

Special Occasions

Whenever the flag is flown at half-mast to honor a deceased dignitary or hero, in the morning it is always raised to the peak first and then lowered to half-mast; in the evening, it is again hoisted to full staff and then lowered completely.

Aboard ship in port, anytime the national ensign is lowered to half-mast, so is the jack.

On Memorial Day, the national ensign is always half-masted when first hoisted at morning colors. At 1200 (noon), a special twenty-one-gun salute is sounded: one every minute until twenty-one shots have been fired to honor those who have given their lives in the defense of our nation. At the conclusion of the firing, the national ensign is hoisted to the peak and flown that way for the remainder of the day. If a twenty-one-gun salute cannot be fired, the ensign is raised to the peak at precisely 1220 (12:20 pm).

During burial at sea, the ensign is kept at half-mast from the beginning of the funeral service until the deceased is committed to the deep.

Shifting Colors

Another custom, less formal than morning or evening colors, yet unique to the sea services, is what we call "shifting colors." As already discussed, the national ensign is flown from the flagstaff at the stern (and the jack is flown from the jackstaff at the bow) when a Navy ship is in port. But when a ship gets underway (no longer moored to a pier or anchored) the national ensign is flown from the gaff (a short angled pole that is higher up and toward the middle of the ship, usually just aft of the main superstructure). When the last line is brought on board, or the anchor is lifted clear of the bottom of the harbor (aweigh), a long whistle blast is blown over the ship's 1MC and the national ensign is hoisted to the gaff while, simultaneously, the national ensign and jack are taken down from the flagstaff and jackstaff respectively. This is all done smartly; the jack and ensign should virtually disappear from the bow and stern while a different ensign leaps to the gaff at the same instant. A ship that does not shift colors smartly will soon have a reputation she does not want.

When a ship returns from sea, the exact opposite procedure takes place. The ensign is taken down from the gaff and another is raised at the stern along with the jack at the bow when the first mooring line is passed to the pier or the anchor touches bottom. All of this is also done smartly, of course, to preserve the ship's reputation as "a taut ship" (efficient and proud).

Underway

Ships at sea do not observe morning or evening colors, but they do fly an ensign at the gaff. The jack is not flown at sea.

When far out at sea and very few other vessels are around, the ensign is often taken down. This is done because a flag flying in the wind suffers a great deal of wear and tear, making it necessary to replace them frequently, and flying the national ensign at times when there is no one around to see it is wasteful.

Dipping

A very old custom of the sea is that merchant ships "salute" naval vessels by dipping their ensigns as they pass by. When a merchant ship of any nation salutes a ship of the U.S. Navy, she lowers her national colors to half-mast. The Navy ship returns the salute by lowering her ensign to half-mast for a few seconds (if the merchant is from a nation that is formally recognized by the U.S. government), then closing it back up. The merchant vessel then raises her ensign back up.

If a naval ship is at anchor or moored to a pier and a passing merchant ship dips her ensign, the salute should be returned as it is at sea—from the flagstaff, where it flies in port, but the jack is not dipped as well.

Naval vessels dip the ensign only to answer a salute; they never salute first. Naval vessels do not dip to one another; they do exchange passing honors instead, if close enough (see below).

Other Flag Display Customs

When displaying the national ensign, there are some customs that are normally practiced, not just in the Navy, but everywhere.

- When the ensign is displayed with other flags, such as service flags (see below) or state flags, it is either positioned in the center and higher than (or in front of) the others or (more often) to the right of the other flags. Imagine a group of flags lined up on a stage as people facing the audience, the national ensign would be placed to the right of the others, so that the ensign would appear to the audience to be to the left of the others.
- If more than one flag is displayed on the same staff, the national ensign goes above the other(s).
- On a stage (such as in an auditorium), the national ensign is placed to the speaker's right. If an ensign is placed off the stage (on the same level as the audience), then the flag is placed to the audience's right (left of the speaker).

- Anytime the national ensign is displayed flat on a wall, the blue field with stars (sometimes called the "canton" or "union") should be seen to the observer's left (the flag's own right). So, for example, if an ensign is displayed on the wall behind a speaker, the stars on the flag will be over the speaker's right shoulder. This applies whether the flag is hung horizontally or vertically.
- In a passageway or lobby the ensign is hung so that the stars are seen to an observer's left (flag's own right) as he or she enters. Again, this applies whether the flag is hung horizontally or vertically.
- In a window, the stars should be seen on the left side of the flag by anyone looking at it from the street.
- If the ensign is suspended over a sidewalk, the stars should be away from the nearest building.
- If the ensign is displayed over the middle of a street, the stars should be to the north on an east-west street, or to the east on a north-south street.
- On an official vehicle, the flag should be displayed on both forward fenders or just the right one (viewed from inside the vehicle; i.e., passenger's side).
- On a casket, the ensign is placed so that the stars are over the left shoulder of the deceased.

There are certain rules for displaying flags that are often ignored. For example, the American flag should not be held parallel to the ground, yet we often see that done at sporting events (often carried by military personnel). Flags technically should not be worn as clothing, yet we see that rule bent or broken frequently. Rather than get too concerned about such things, the overarching rule is *intent*. If a giant flag is carried parallel to the field during halftime at the Super Bowl, or someone wears a sweater that looks for all the world like an American flag, these are meant to be expressions of patriotism, and that trumps technical rules.

Other Flags

Many other flags besides the national ensign are used in the Navy. There are flags called signal flags that represent the letters of the alphabet, numbers, and some specialized meanings that are used by ships to communicate ceremonially and tactically (even today in the age of radio communications, because signal flags do not broadcast a signal that can be detected by an enemy). There are jacks as previously discussed. And there are also other traditional flags used for a variety of purposes as discussed below.

Commission Pennant

The commission pennant is long and narrow, with seven white stars in one row on a blue field and the rest of the pennant divided lengthwise, red on top and white below.

The commission pennant flies, day and night, from the time a ship is commissioned until she is decommissioned (in other words, while she is in service as a U.S. Navy ship), except when a personal flag or command pennant is flying instead (as explained below). One other exception is that a Navy hospital ship flies a Red Cross flag instead of a commission pennant.

The commission pennant is hoisted at the after truck (top of the mast closest to the rear of the vessel) or, on board a mastless ship, at the highest and most conspicuous point available.

A commission pennant is also flown from the bow of a boat if the commanding officer is embarked (riding in the boat) to make an official visit.

The commission pennant is not a personal flag, but sometimes it is regarded as the personal symbol of the commanding officer. Along with the national ensign and the jack, it is half-masted on the death of the ship's commanding officer. When a ship is decommissioned, it is the custom for the commanding officer at the time of decommissioning to keep the pennant.

Personal Flags

You will frequently hear the terms "flag officer" or "flag rank." These terms refer to admirals (generals in other services) and have come about because officers in paygrades O-7 through O-10 in the services have special flags that accompany them wherever they go in an official capacity. For Navy admirals the flags are blue and white and have the same number of stars that these officers wear on their collars and on their shoulder boards: one star for a rear admiral (lower half), two stars for a rear admiral (upper half), three stars for a vice admiral, and four stars for an admiral. For staff corps officers (supply, medical, etc.) they have blue stars on a white field; for line officers they have white stars on a blue field; because the CNO is senior to all staff and line officers, the background of his or her flag is half blue and half white. These flags are flown on the fenders of their official cars, in front of their headquarters ashore, and from the main truck (top of the tallest mast) on board ships in which they are embarked.

The commission pennant and the personal flag of an admiral are never flown at the same time, so if a vice admiral boards a ship, the commission pennant is hauled down from the after truck and the admiral's three-star flag is hoisted instead at the main truck. The admiral's personal flag remains flying for as long as the admiral is officially embarked, even if he or she leaves the ship for a period of less than seventy-two hours.

Some very high-ranking officials, such as the SECNAV and the CNO, have their own specially designed personal flags that are flown in the same manner as the starred flags of admirals.

Command Pennants

Officers who are not admirals but have command of more than one ship or a number of aircraft rate a command pennant. These pennants are flown similarly to admirals' personal flags. When commanding a force, flotilla, squadron, cruiser-destroyer group, or aircraft wing, the officer rates a "broad command" pennant, which is white with narrow blue stripes along the top and bottom. An officer in command of any other unit, such as an aircraft squadron, flies a "burgee command" pennant, which is white with narrow red stripes top and bottom.

Absence Indicators

When a commanding officer or any flag officer is temporarily absent from a ship, an absentee pennant is flown. When the admiral or unit commander whose personal flag or command pennant is flying departs the ship for a period less than seventy-two hours, his or her absence is indicated by hoisting a special pennant called the first substitute to the starboard yardarm. The second substitute, flown from the *port* yardarm, indicates that the admiral's chief of staff is absent. The third substitute, also flown from the port yardarm, indicates the absence of the ship's commanding officer. The fourth substitute flying from the starboard yardarm means that a civil or military official who is officially embarked (such as the SECNAV) is absent. It is flown from the starboard yardarm.

Unit Awards

Ships and shore commands also fly flags or pennants representing special awards the ship or command has earned. These include such things as a small triangular pennant with a ball in the center (unofficially called "the meatball") for battle efficiency, the Presidential Unit Citation (PUC), Navy Unit Commendation (NUC), or Meritorious Unit Commendation (MUC).

Service Flags

Each of the services has its own official flag that is displayed in offices, on stages, at special events, and so on. When more than one service flag is displayed (such as at a joint command or a ceremony involving all or several of the services), they should be displayed in a special order (and it may not be the one you would think of).

The order of precedence for service flags is:

Army

Marines Corps

Navy

Air Force

Coast Guard

For example, if you imagine a group of flags lined up on a stage as if they were people facing the audience, the national ensign would be placed to the right of the others, with the Army flag next, and then the Marine Corps flag and so on. From the audience's perspective the ensign would appear to be to the left of the others.

This precedence may seem odd, but it is based upon a lot of confusing history involving which service began first officially. As often happens, historians have changed their minds as to what happened when and what event was more significant. A detailed description of all of this is beyond the scope of what we are trying to accomplish here. Just be sure—if you are ever tasked with displaying more than one service flag—that you put them in the order shown above. If anyone asks you why that order, just say "because."

If, as is usually the case, a national ensign is displayed with the service flags, it is of course at the head of the line.

The higher precedence flag goes to the right of the next higher precedence. This is accomplished by thinking of the flags as facing in a particular direction. For example, if you were setting up all of the service flags (and the national ensign) on a stage, they would be set up with the Air Force flag to the right of Coast Guard flag, the Navy flag to the right of the Air Force flag, and so on up to the ensign. This means that the audience would see the flags in the following order: national ensign, Army, Marine Corps, Navy, Air Force, and Coast Guard. In other words, the higher precedence flags are to the right of the others *as they face the audience*, but the audience would see them to the left.

In a parade, the color guard would be arranged so that the national ensign is to the right of all the other flags, as they faced in the direction of the parade. Therefore, the order would be like this:

Coast Guard
Air Force
Navy   *Direction of march* →
Marine Corps
Army
national ensign

Battle Streamers

You might notice a lot of colored streamers attached at the top of the Navy flag (or on the flags of other services as well). These are *battle streamers*, and each one commemorates a major battle, campaign, or war in our nation's history. They serve as a visual reminder of the service and sacrifice rendered in those momentous events.

The U.S. Army was the first to adopt battle streamers in 1920. The Marine Corps followed suit in 1939, the Air Force in 1956, the Coast Guard in 1968, and the Navy in 1971. How these streamers are assigned differs from service to service. For example, the Army and Air Force assign a separate streamer for each important action; consequently, the Army has more than 150 streamers and the Air Force has more than sixty. The Navy and Marine Corps often use one streamer for an entire war or campaign and add embroidered stars to the streamers to represent individual actions. Currently, the Navy has twenty-nine battle streamers.

The Navy's streamers are 3 feet long and 23/4 inches wide, and each one is unique in its combination of colors to represent an individual war, campaign, or theater of operations. The embroidered stars represent individual battles or specific operations that rate special recognition. One battle or operation is represented by one bronze star. To save space, five battles or operations are represented by one silver star. On the battle streamer for the War of 1812, for example, fourteen separate battles are represented by two silver stars and four bronze stars. And the three battles of the Quasi-War with France are recognized by three bronze stars. World War II was such a vast war that it is represented by a series of streamers, each representing a different area (theater) of operations (Asiatic-Pacific Theater, European-African-Middle Eastern Campaign, etc.).

Taken together, these battle streamers provide a colorful representation of the operational history of the U.S. Navy.

OTHER STRANGE HABITS AND PRACTICES

There are countless other customs and traditions in the Navy. Unfortunately, there is no single official source for learning about them all. Some are covered in Navy Regulations, but not all. Some are prescribed in separate instructions, and many are not officially addressed at all, yet are practiced out of simple tradition. One excellent source of information is *Naval Ceremonies, Customs, and Traditions*, now in its sixth edition, by Royal W. Connell and William P. Mack.

Though we cannot cover them all, below are some brief descriptions of some you may encounter

"Yes" versus "Aye Aye"

In the chapter on "Military Titles" there is a detailed discussion of how to address military people by rank. What is not covered there is how Navy people are expected to respond.

It will probably come as no great surprise that when officers in the Navy ask someone junior to them (another officer or an enlisted person) a question that can be answered with a simple "yes" or "no," it is customary for the junior to answer, "Yes, sir," "Yes, ma'am," "No, sir," or "No, ma'am," whichever is appropriate. If the answer requires more than a simple yes or no, "sir" or "ma'am" is still added to the answer, such as, "The boat is ready for launching, sir." But if an officer gives an order (in other words, tells someone to do something), the proper answer in the Navy is, "Aye, aye, sir," or "Aye, aye, ma'am." This means that the responder has heard the order, understands it, and will carry it out.

When a senior wants to indicate that she or he has heard and understood a report from a junior, she or he will answer, "Very well." A junior should never say, "Very well," to a senior. Observe the following conversation:

Lieutenant Washington: "Seaman Nelson, is the boat ready for launching?"

Seaman Nelson:	"Yes, sir."
Lieutenant Washington:	"Excellent. Make certain there are enough life jackets on board."
Seaman Nelson:	"Aye, aye, sir."
(Nelson checks.)	
Seaman Nelson:	"There are seven life jackets on board, sir."
Lieutenant Washington:	"Very well."

Chief petty officers are not addressed as "sir" or "ma'am," but they are customarily responded to as "Chief" (if an E-7), "Senior" (if an E-8), and a variety of titles (such as "Master Chief" or "Force") if they are E-9s. (See chapter 1, "Military Titles," for a more detailed explanation.) Both juniors and seniors will say "Yes, Chief," or "No, Senior," and so on when responding to these individuals.

"Respectfully" versus "Very Respectfully"

Though careers are not ended over such matters, a potentially embarrassing situation can occur if one does not know the "secret code" of correspondence closings. A senior customarily closes a written communication (most often a memo or e-mail) to a junior with the word "Respectfully," whereas juniors (when communicating with their seniors) are expected to add the word "very," so that the closing becomes "Very Respectfully."

Abbreviating the closing is also acceptable, but the aforementioned custom is not suspended, so that seniors may merely use an "R" for "Respectfully" and juniors will use "V/R." In very informal communications (such as an e-mail), the letters may be lowercased, but the "r" and "v/r" custom persists.

As a civilian, it is doubtful that anyone will expect you to adhere to this "code," so "sincerely" or whatever you are comfortable with is probably fine, but it helps to be aware of these practices. It is possible that, as the integration of civilians into the Total Force concept continues and matures, this kind of thing may be expected of you as well.

Attention on Deck

Whenever important visitors, the captain, or other senior officers approach an area or enter a compartment (room) where there is a gathering of personnel, the first person to see her or him approaching will call out, "Attention on deck." All present should immediately come to attention and remain that way until the senior person present gives the command, "Carry on." This courtesy applies (even the use of the word "deck") both on and off ships. As a civilian, you are not expected to call out the command, but you should "do as the Romans do" and come to your feet with the others and wait for "carry on" before sitting.

Gangway

You may hear the command "Gangway!" given by anyone who observes an officer or important official approaching where passage is blocked. Everyone blocking the passage should immediately move out of the way. As a civilian you are not expected to call out the command (although there is nothing wrong with your doing so), but you should certainly get out of the way once you hear it.

Customarily, no one should make the call for themselves, and enlisted personnel do not clear a passage for other enlisted personnel in this way. "Coming through" is appropriate to say in these circumstances instead.

You may sometimes hear "make a hole" used when someone wants to clear a path. It is *not* the correct term to use, but the reality is that it is frequently used (incorrectly) and you would be well advised to react to it to avoid being trampled or to avoid the embarrassment of impeding others who are trying to get through, probably for good reason.

Boats and Vehicles

In the Navy, the customary rule for getting in and out of boats and vehicles (and elevators for that matter) is *seniors in last and out first*. Unless otherwise directed, those junior in rank will board boats before their seniors and will debark after. The same is generally true with vehicles (particularly with a large group, such as when boarding a bus), although practicality or other considerations sometimes overrule this custom. A junior who is driving a senior will usually open and close the door for the senior, so the customary order is obviously reversed in that case.

In boats, seniors sit aft of juniors. If you embark in a boat, avoid stepping on deck areas that are heavily varnished for appearance's sake.

The Sailor in charge of a boat is traditionally called a coxswain. Should you have to say that unusual word (as in, "Coxswain, are we going to sink?"), you should know that it is pronounced *cock-sun* (not *cocks-swane*).

Salty Language

We refer here not to the scatological aspects of nautical communications (one could probably write a whole other book on that subject) but to the anomalies in pronunciation and spelling that one encounters when communicating with those of the nautical persuasion.

Strange Pronunciations

As with coxswain (explained above), there are a number of words commonly used in the Navy that are not pronounced the way they appear. The reasons for this vary but mostly are because many of our Navy terms came to us via the Royal (British) Navy. Though we will not lose a major battle if you mispronounce these words, you will avoid some personal embarrassment if you use the traditional pronunciation rather than the phonetic.

Like coxswain, the word "boatswain" is pronounced *BOH-sun* (emphasis on the first syllable), not *boat-swane*. You will also encounter the more correct term "boatswain's mate"; this is also pronounced as *BOH-sun's mate*, although you will often hear the second "s" dropped, so that it sounds more like *BOH-sun mate*.

Many of the Navy's ratings end with "man," such as "Damage Controlman," "Legalman," and "Hospital Corpsman." The "man" part is not emphasized and is in fact de-emphasized, so that these words become more like *LEGAL-mun* than *legal-MAN*.

And the "s" is not pronounced in "Corpsman," so that it becomes *CORE-mun*.

"Forecastle" is another such term that requires special care and feeding. The proper way to say this one is *FOHK-sul*, not as it appears.

Do not emphasize the "board" in "starboard"; it should be pronounced more like *STAR-bird*.

"Halyard" (line used to haul up a flag, pennant, etc.) is pronounced *HAL-yird*, not *HALL-yard*.

When talking about the front end of a vessel, "bow" is pronounced the same as when a performer takes a bow, not as in a bow tie.

A common knot used by sailors is a "bowline" and it is pronounced *BOH-lin*, but a line running from the front of a ship is called a bow line (pronounced as in a performer's bow; and line is said as in an actor's line in a script).

A person who uses a line to measure the depth of the water is called a leads-man and it is pronounced *LEDS-mun*, not *LEEDS-mun*.

To avoid confusing the similar sounds (particularly on a radio or tele-phone) of "five" and "nine," the latter is often pronounced *niner*.

The word "mast" is pronounced more like *mist* when accompanied by a modifier, so that "foremast" becomes *FOR-mist*, "main mast" becomes *MAIN-mist*, and so on. However, "mast" without any modifiers does not change, so that you would *not* say "The antenna is mounted on that *mist*."

Nautical purists will insist that whereas the "lee" side of the ship (the downwind side) is pronounced exactly as you would expect, the word "lee-ward" (as in, "Make sure you maneuver the ship so that the man overboard winds up on the leeward rather than the windward side") is pronounced *LEW-word*. Real purists also say the word "windward" as *WIND-erd*, and the word "tackle" (as in a "block and tackle rig") is said as *TAY-cul*. These purist practices seem to be fading (for better or worse is debatable), but there are many who still use them, so it does not hurt to be aware.

Pronunciations You Will Hear But Should Not Mimic

There are also some odd but widespread mispronunciations in the Navy. For example, when the legendary Vice Admiral John Bulkeley was still with us (his career spanned many decades from his World War II rescue of General MacArthur from the Philippines to his long-term tenure as head of the Insurv Board), he was almost universally incorrectly called "Admiral Buckley."

The plug that fits into the barrel of a gun is a "tompion" (tampion is an acceptable alternative spelling) but for some reason, Sailors more often call it the "tompkin."

A large rope used in towing or mooring is a "hawser" and is properly pro-nounced *HAW-zer*, yet you will often hear Sailors pronounce it *HOW-zer*.

A "console" is all too often incorrectly called a *KOWN-sul*.

Why these things come to pass and why they become the rule more than the exception is one of those many mysteries of human nature.

Odd Spellings

There are also some strange *spellings* in the Navy. For example, the phonetic word for the letter "A" is spelled *Alfa*, not *Alpha*. The phonetic letter for "J" is

Juliett, not *Juliet*. And although Webster's dictionary recognizes two spellings of the drink "whiskey" (also "whisky"), when used as the phonetic for "W" in the military, it is spelled *Whiskey*, not *Whisky*.

When dealing with weapons, a common error to be avoided is to confuse the word *ordinance* with *ordnance*. The former has nothing to do with weapons. Also, the detonating device in a projectile, bomb, or torpedo is a *fuze*, not a *fuse*.

Music

Besides the national anthem, there is quite a bit of music associated with the Navy. Various marches (including an "Admiral's March" played to honor flag officers, much as "Hail to the Chief" is played to honor the President) are played at many official occasions, sea chanties (shanties) and other seafaring songs are a part of the Navy culture, and "ruffles and flourishes" are played (by all military services) as a way of honoring dignitaries at very formal occasions.

Two pieces—"Anchor's Aweigh" and "The Navy Hymn"—stand out among the many associated with the Navy.

"Anchor's Aweigh"

This is the Navy's "fight song," anthem, march, and a few other things. Please note how it is spelled—refrain from the common error of writing it as "Anchors Away" or various other variations that too often appear. It refers to a ship's anchor being "aweigh" (lifted off the bottom).

Originally meant to inspire Naval Academy midshipmen at the annual Army-Navy game, the music was composed in 1906 by Lieutenant Charles A. Zimmerman, the bandmaster at the U.S. Naval Academy, and the original lyrics were written by Midshipman Alfred Hart Miles. Navy beat Army by a score of 10–0 that year and the rest is history, as they say.

The original football-oriented lyrics have been joined over the years by numerous variations, but the most recent version—meant to be all-service inclusive—is as follows:

Stand, Navy, out to sea, fight, our battle cry;
We'll never change our course, so vicious foe steer shy.
Roll out the TNT, anchor's aweigh. Sail on to victory
And sink their bones to Davy Jones, Hooray!

Anchor's aweigh, my boys, anchor's aweigh.
Farewell to foreign shores, we sail at break of day, of day.
Through our last night on shore, drink to the foam,
Until we meet once more, here's wishing you a happy voyage home.

"The Navy Hymn"

The song known to U.S. Navy men and women as "The Navy Hymn" (also sometimes called "Eternal Father") is a musical benediction that long has had a special appeal to seafaring men and women, whether they are religious or not. Reverend William Whiting wrote the original words to the hymn around 1860 after surviving a furious storm in the Mediterranean. In 1861, the words were adapted to music by another English clergyman, the Reverend John B. Dykes, who had originally written the music as "Melita" (ancient name for the Mediterranean island of Malta).

The haunting melody has been teamed with countless versions of lyrics—many of them adapted to aviation, Marines, and so on—but the most commonly used verse is as follows:

Eternal Father, Strong to save,
Whose arm hath bound the restless wave,
Who bid'st the mighty Ocean deep
Its own appointed limits keep;
O hear us when we cry to thee,
For those in peril on the sea.

"Eternal Father" was the favorite hymn of President Franklin D. Roosevelt and was sung at his funeral at Hyde Park, New York, in April 1945. Roosevelt had served as Assistant Secretary of the Navy and felt a strong personal attachment to the Navy. This hymn was also played as President John F. Kennedy's casket was carried up the steps of the U.S. Capitol to lie in state after his assassination in November 1963; he also had served in the Navy.

Frocking

The practice that we currently refer to as "frocking" has been in common usage in one form or another throughout the Navy's history. This occurs when a Sailor is permitted by higher authority to assume the uniform of a higher rank, usually without the accompanying higher pay, when appointed by proper authority to assume the duties and responsibilities of that rank prior to actual promotion. These appointments were sometimes temporary as when it was necessary rapidly to swell the ranks during wartime, especially in the Civil War. Other instances concerned a commander's need to fill an unforeseen vacancy for which there was no one of equal rank. In this case, a junior who was qualified for promotion would assume the uniform and duties pending approval by the Navy Department. Today, a Sailor who has been selected for promotion but has not reached the official date of promotion may be "frocked"; that is, permitted to wear the rank (but not receive the pay) until he or she reaches the actual promotion date (at which time the pay begins as well). This is not a "given" but is permitted on a case-by-case basis.

The origins of the term are not actually known, and there are several possible explanations. The most likely one is that the early undress uniform for a midshipman was a short coat whereas that for a lieutenant was a longer "frock coat." When a midshipman was appointed to act in the capacity of a lieutenant, he wore the uniform of the latter and was said to be "frocked."

There is a formal military instruction (DOD Directive 1334.2) that details current policy on frocking.

Ship-Related Customs

As one might expect, there are a number of customs related to ships. Elaborate ceremonies take place when ships are launched, commissioned, and decommissioned. If you are invited to attend one of these, you should go if at all possible; it will give you an appreciation for the Navy's heritage and traditions in a special way. Whether you actually go to sea on a Navy ship or just have occasional contact with them, it helps to be aware of some of the other strange customs you might encounter.

Manning the Rail

If you are present when a Navy ship returns from a long deployment, you may see Sailors formally lined up at regular intervals along the ship's rails on the main deck and up on the superstructure wherever there is a visible place to stand. This is called (appropriately enough) "manning the rail" and is an offshoot from an old custom.

In the days of sail, the captain of a ship could honor a distinguished person by directing his crew to "man the yards." This entailed having the crew members stand evenly spaced on all the ship's yards (the crossbars on the masts from which the sails were suspended) and give three cheers. With the disappearance of sails (and most of the yards) the custom has been modified to its present form. It is still used to render honors to a head of state or to a member of a reigning royal family, but it is more often practiced now as a means of honoring the family members who were left behind to await the ship's return from a long deployment. It is a sight to behold.

Dressing Ship

Commissioned ships are "dressed" on national holidays and "full-dressed" on Washington's Birthday and Independence Day. When a ship is dressed, the national ensign is flown from the flagstaff as usual but others are added to each masthead as well.

When a ship is full-dressed, in addition to the added ensigns, a "rainbow" of signal flags is displayed from bow to stern over the mastheads, or as nearly so as the construction of the ship permits. These flags are displayed in a prescribed order that appears random and does not spell anything or have any additional meaning. This prescribed order ensures that a creative signalman does not have the option of spelling "Hello, Aunt Gertrude," or worse, when rigging this display.

Ships are only dressed and full-dressed in port, never underway and only from 0800 to sunset.

The rainbow of signal flags is also employed at other times, such as when visiting a port other than the ship's home port. It is employed to make the ship more attractive or more inviting when visitors are expected.

Passing Honors

If you are embarked in a Navy ship that is underway, you may encounter the practice of an old tradition of the sea that occurs when naval vessels pass within close proximity of one another. (The term "naval vessel" as used here means any U.S. Navy or Coast Guard vessel and foreign warships belonging to nations formally recognized by the United States.) The prescribed distance for what is called "passing honors" is six hundred yards for ships and four hundred yards for boats. This means that this will generally only occur when ships are entering or leaving port, where approach channels are relatively narrow causing vessels to pass within close proximity; vessels are unlikely to pass that close in the open sea. When your ship is entering or leaving port or in the rare occurrence of a close encounter at sea, if you are on deck where you can be seen by personnel on a passing naval vessel, you should be prepared for the implementation of passing honors.

The junior vessel always initiates the honors. "Junior" is determined by the relative seniority of the vessel's commanding officers or by the seniority of an embarked official (such as an admiral or a civilian dignitary). If you are a Deputy Assistant Secretary of the Navy, for example, these passing honors are probably for you.

The process goes like this. The junior vessel initiates the passing honors by passing the word "attention to port" or "attention to starboard" depending upon which side of the vessel the senior ship is on. All members of the crew who are outside on the weather decks will stop what they are doing (unless their work is safety-related and it would be dangerous for them to stop) and face the direction indicated at attention. (From another chapter in this book, you should be able to determine which side is port and which is starboard, but if you are in doubt, look around and see what others around you are doing. If you are alone, look out to see if you see a ship in close proximity; if you do not, chances are, you are on the wrong side of the ship and do not have to worry about what to do.) The vessel being honored will likewise call its crew to attention, facing toward the junior vessel.

Next, the words "hand salute" are passed on the junior vessel and the hand salute is rendered by all persons on deck. This also is returned by the senior vessel. "Two" (the command for ending the salute) is then passed by the *senior* vessel, followed by the junior. During this saluting process, you should simply remain standing respectfully, facing the other vessel.

Once the vessels are clear, "carry on" is sounded and everything returns to normal routine.

Sometimes the entire process is accomplished using a series of whistle signals blown with a police-type whistle without any accompanying verbal commands.* In that case, you should merely observe what crew members around

---

*If you are more comfortable knowing what the signals mean, they are as follows: One whistle indicates attention to starboard; two indicates attention to port. Subsequent commands are one whistle for hand salute, two for ending the salute, and three for carrying on.

you are doing and use their actions as your cues. The important thing is that you are not seen sitting or lounging about or doing anything that might be construed as less than respectful.

Passing honors are also rendered by vessels passing the Arizona Memorial in Pearl Harbor and Mount Vernon on the Potomac River.

Tending the Side (Side Boys)

There is another shipboard custom that you may encounter (in fact, it is occasionally mimicked ashore, especially during retirement ceremonies) is the ancient nautical custom of tending the side. This special ceremony is invoked when a visitor who rates special attention, such as a senior officer or a high-ranking civilian official, comes aboard or departs. To carry out this honorary ceremony, the Boatswain's Mate of the Watch "pipes the visitor aboard" by blowing a special call on a boatswain's pipe while an even number (depending upon the rank of the officer or official) of Sailors line up at attention on either side of the gangway (entrance to the ship) forming a human corridor for the dignitary to pass through. The Sailors salute on the first note of the pipe, holding their salute as the person being honored walks through the human passageway, and finishing the salute together on the last note of the boatswain's pipe.

The Sailors assigned to this duty are particularly smart in appearance and well groomed, with brightly polished shoes and immaculate uniforms. They are traditionally referred to as "side boys," no matter what their gender.

An extension of the custom is sometimes carried out on the flight deck of an aircraft when a dignitary arrives by airplane or helicopter. In this case, the side boys are mustered wearing the different colored jerseys they wear when working on the flight deck to indicate what their roles are (plane handlers wear yellow, fueling personnel wear purple, weapons handlers use red, etc.). In this event, the side boys are called "rainbow side boys."

Just in case you are the subject of this honorary custom and are concerned about it going to your head, you might consider the origin of the custom. In days long gone, there was no dignified way for an official to scramble up a ladder from a boat to the main deck of a ship, so a "boatswain's chair" was rigged over the side to hoist the dignitary aboard. Legend has it that the some of these honored persons were a bit corpulent from too many years of privileged living so more side boys were required. A natural formula developed that said, "The more senior, the more corpulent, the more boys required."

Bells and Gongs

In chapter 7, "Shipboard," we will discuss the use of ship's bells as a traditional means of telling time, but there are other bells and bell-like sounds aboard ship.

ARRIVALS AND DEPARTURES

If you are aboard a Navy ship and suddenly hear a series of bells or gongs ringing over the 1MC and the words, "Reagan arriving," do not worry that you have suddenly traveled through time or that the ghost of the former President is about to make an appearance. This is part of an old custom that actually serves a purpose.

To begin with, you should know that traditionally the captain of a ship takes on the name of that ship while in command. So the commanding officer of USS *Leyte Gulf* becomes "*Leyte Gulf*" in certain circumstances.

Further, it is important for certain members of the crew to know when the captain is departing or arriving—the command duty officer (officer in charge of the ship when the captain is not aboard) is one excellent example.

So, when the captain departs or arrives on board his or her own ship, a series of bells (the number is always even and is determined by her or his actual rank) is sounded for the entire crew to hear, and the words "*Leyte Gulf*, arriving" (or departing) are passed as well. When the captain actually steps on (or off) the ship, another single bell is struck.

This practice carries over to those occasions when the commanding officer of one ship departs his or her own ship and visits another. For example, if the captain of USS *Leyte Gulf* left her own ship to visit with the captain of USS *Patriot*, on board *Leyte Gulf* you would hear four bells (because she is a captain [O-6] by rank) followed by the words "*Leyte Gulf* departing." After a few minutes (to allow time for the command duty officer or anyone else who might need to see the captain before her departure to get to the quarterdeck before the captain actually departed), you would then hear a single bell as the captain actually stepped off the ship. Later aboard *Patriot*, as the captain of *Leyte Gulf* approached, you would hear four bells and the words "*Leyte Gulf* arriving" sounded, followed by a single bell as she actually stepped aboard *Patriot*. If the two captains decided to leave *Patriot* to go to a meeting ashore, you would hear the following aboard *Patriot*:

Four bells.
"*Leyte Gulf* departing."

Two bells (because the captain of *Patriot* is a lieutenant commander [O-5] by rank).

"*Patriot* departing."

A single bell (as *Leyte Gulf*'s captain steps off *Patriot*).

Another single bell (as *Patriot*'s captain steps off).

This practice does not just apply to commanding officers of ships; others in command are treated the same way. For example, if the vice admiral in command of the 7th Fleet were to visit USS *Leyte Gulf*, you would hear eight bells (because he or she is a vice admiral by rank), followed by the words "7th Fleet arriving."

NAVIGATIONAL BELLS AND GONGS

Although not a custom, there is a continuing legal requirement of the Navigational Rules of the Road that goes back many centuries. Ships are required to ring a bell forward (and a gong aft on larger vessels) when at anchor during low-visibility conditions. By making these sounds, vessels help each other avoid collisions. Vessels that are underway sound fog signals on their ship's whistles or on special fog horns. So with these different signals (bells and gongs versus whistles and horns) it is possible to not only know that there are vessels present in the vicinity but also whether they are fixed in place (anchored) or moving.

BOAT GONGS

When a ship is at anchor, a series of boat gongs are often sounded over the ship's 1MC to alert personnel on board to the upcoming departure of boats. This allows one to head for the quarterdeck in time to board the boat before it departs. The sequence is as follows:

three gongs	boat will depart in ten minutes
two gongs	boat will depart in five minutes
one gong	boat will depart in one minute

Pollywogs, Golden Dragons, Blue Noses, and the Like

There are a number of customs in the Navy that are of absolutely no practical use, yet they are important in a kind of spiritual way. These are initiations into certain "realms" that occur when individuals achieve specific geographic or mission milestones. They are serious in that they tie today's Sailors to those who have sailed the seven seas before them, passing on a legacy and a kind of kindred "soulship" that spans time and nationality. But they are touched with humor as well, providing an opportunity for all to feel momentarily equal as fellow humans, no matter what rank or age the participants may be. An admiral is subject to the whims of Neptune as readily as is the youngest neophyte seaman, and so all are expected to prove themselves worthy of the strange new titles they are called upon by happenstance to earn.

There is actually a formal Navy directive (SECNAVINST 1610.2) that details the Navy's policy concerning such military functions that involve initiations or other similar ceremonies. Though it recognizes these events as important to tradition, it also is designed to ensure that they do not get out of hand, as things of this nature (such as fraternity initiations) can do if not properly supervised and controlled.

CROSSING THE LINE

The granddaddy of all seagoing initiations is the crossing the line ceremony. Its origins date from a more superstitious era when man appeared to be at the mercy of the god of the sea. It is practiced when a ship crosses the equator. Sailors who have not previously crossed that unique line are known as *pollywogs* and must undergo the initiation process to become *shellbacks* or "fit subjects of King Neptune." No one is exempt from going through the initiation (unless they have strong feelings to the contrary and specifically opt out of the festivities)—even President Franklin D. Roosevelt participated in a crossing the line ceremony in 1936 while embarked in USS *Indianapolis*.

The crossing the line ceremony is a colorful tradition full of ritual and all sorts of bizarre goings-on. Those who are already shellbacks conduct the ceremony and play the role of members of King Neptune's court. Pollywogs must go before the court and answer for charges of violating the king's realm by enduring certain rites of passage that, once completed, result in their being awarded shellback certificates as proof of their new status, so that the newly appointed shellbacks need never repeat the initiation.

GOLDEN DRAGON

Similar to crossing the line is the *Order of the Golden Dragon*. This distinction marks the crossing of the 180th meridian, the international date line.

Since World War II, the U.S. Navy has deployed to the Far East so frequently that few initiation ceremonies are still actually held, but those who cross for the first time are still eligible for the certificate that recognizes this "achievement."

GOLDEN SHELLBACK

One who crosses the intersection of the 180th meridian and the equator is granted this special status and an appropriate certificate.

EMERALD SHELLBACK

This is one who crosses the intersection of the Greenwich meridian (0ring longitude) at the equator, which occurs in the Gulf of Guinea off the west coast of Africa.

BLUE NOSE

This distinction marks the crossing of the Arctic Circle, where the Sailor is said to have entered the *Northern Domain of the Polar Bear*. Arctic Circle certificates have a long seafaring tradition nearly as old as shellbacks, going back as far as the Middle Ages. Today, King Polar Bear is piped aboard at the limit of his domain and lets his wrath be known to the uninitiated. The certificate denotes the individual as a member of the *Royal Order of the Blue Noses*.

RED NOSE

This distinction marks the crossing of the Antarctic Circle to enter the *Royal Domain of the Emperor Penguin*. His Imperial Majesty inducts the individual as a "Frozen Stiff." The bearer of this certificate is entitled "to all of the privileges of this frozen realm of blizzards, including freezing, shivering, starving, and any other privileged miseries that can possibly be extended during his stay in this land of answer to a well digger's dream."

MOSSBACK

This distinction marks the passage around Cape Horn at the southern tip of South America. Holders of this certificate are an exclusive brotherhood because the Panama Canal makes such a journey rarely necessary. Because of the passage's reputation for stormy weather, the certificate grants the holder the right to spit into the wind, if he or she wants to risk it.

ORDER OF THE DITCH

This distinction is granted to those who transit the Panama Canal. The certificate traditionally reads, "By canal, the distance between the two oceans is about 50 miles. It would be 7,837 miles around South America."

SPECIAL CRUISES AND MILESTONES

The *Order of the Spanish Main* is granted for those who cruise the Caribbean Sea. Service in the Persian Gulf is commemorated with a certificate for a *Persian Excursion*, which inducts the sailor into the *Mystic Society of One Thousand and One Nights*. Those who have sailed the Black Sea are recognized for having entered the *Realm of the Czars*.

Special certificates are sometimes presented to submarine Sailors to commemorate their first dive, last dive, or other special occasions. "Deep Dive" diplomas are also sometimes given to VIPs and family members of sailors to commemorate their visit to a submarine.

Milestones in a naval aviator's career often rate notice in the form of certificates. Naval aviators join the *Century Club* when they make their hundredth carrier landing and become *Double Centurions* at two hundred landings. Some aviators who flew over Vietnam were inducted into the *200 Mission Club* and the *300 Mission Club*. This tradition has continued in subsequent conflicts.

Another "century club" is an organization of Florida-based hurricane hunters who have flown through winds of one hundred miles per hour or more. Members of this *Not So Ancient Order of the Hurriphooners* receive a scroll bearing the legend: "At wave-level height, this member has battled forces of *Neptunus Rex* and aerial elements of the *Chief High Gremlin* to a standstill." It is signed by the "Most Exalted Hurriphoon Hunter and the High Hurriphoon Cloud Sniffer."

DUBIOUS HONORS

Anyone who has ever been in conditions of potential danger and great stress will tell you that humor is the best antidote. It is in recognition of that maxim that certain other "honors" are sometimes bestowed.

High-Floating Hook-Bouncing Barrier Crashers are those carrier pilots who had to use the emergency barrier to land. The *Goldfish Club* is for those who ditch at sea. If these unfortunate individuals spend more than twenty-four hours on a life raft, they become *Sea Squatters*. Anyone who has made an unscheduled parachute jump from a disabled plane is eligible for membership in the *Caterpillar Club*. Members wear (unofficially) a gold caterpillar pin with jeweled eyes, determined by the circumstances of the jump; for example, ruby eyes show that the wearer has survived a midair collision.

FINAL THOUGHTS

Some of the customs brought about by centuries of seafaring tradition are quite impressive, such as when a ship returns from a long deployment with her rail manned. Some of them are rather silly, such as those initiations and certificates that recognize certain significant but unimportant rites of passage. Some conflict with (and take precedence over) common social practices in civilized society that are gender based, such as when a woman (lieutenant) follows a man (admiral) through a doorway.

But these customs are part of the cement that holds the Navy together, that form a special link with the past and reap the benefits of humor and a bit of "club" pride, while keeping alive traditions that are, in some cases, thousands of years old.

QUICKREFS

- Saluting is a custom with unclear origins that is frequently used by all the armed forces, but is not practiced identically in all the services.
- Sailors generally salute only outdoors and when covered (wearing a cap); though the right hand is preferred, they can use their left hand under special circumstances.
- Soldiers and Airmen may salute uncovered, indoors as well as outdoors (but not left-handed).
- Special circumstances and joint commands can overrule these practices.
- Enlisted personnel salute officers; junior officers salute more senior officers; enlisted personnel do not normally salute one another.
- Personnel from different services (and those of foreign nations) salute one another, generally using the above rules.
- There are special rules for saluting while armed.
- Salutes are *not* exchanged when inappropriate, such as when working, in combat, on a bus, while eating, when guarding prisoners, and so on.
- Gun salutes are rendered to honor dignitaries. The number of guns fired is always an odd number and depends upon the rank or position of the person being honored. The President of the United States receives the maximum number (twenty-one), the Secretary of the Navy rates nineteen, admirals seventeen, vice admirals fifteen, and so on.

- In the Navy, the American flag is often called the "national ensign" or simply "the colors."
- Personnel in uniform salute the national ensign and people in civilian clothes should stand at attention with right hand or hat (if wearing one) over their hearts
  —when the national anthem is being played,
  —when the flag passes by in a parade or similar situation, and
  —during morning colors at 0800 (8 am) and evening colors at sunset.

- When in port, Navy ships display the colors at the *stern* (on the flagstaff) and the *jack* (currently the "Dont Tread on Me" flag) at the bow (on the jackstaff). When underway, Navy ships *shift colors* to the gaff (a small pole high up, usually just aft of the main superstructure) and the jack is put away.
- Merchant vessels often *dip* (lower halfway down) their ensigns (national seagoing flags) to passing warships; Navy ships always answer (but never initiate) these "salutes," unless the nation is not formally recognized by the U.S. government. Warships do not dip to one another.
- There are specific rules for properly displaying the national ensign that cover virtually every situation. Two general rules apply.
- When displayed with other flags, the national ensign is placed at its own right. For example, if you imagine a group of flags lined up on a stage as people facing the audience, the national ensign would be placed to the right of the others, so that the ensign would appear to the audience to be to the left of the others.
- The other general rule is that the part of the flag that is a blue field with stars (also called the "union" or "canton") is displayed to the flag's own right as well. For example, the flag would be displayed on a wall behind a speaker so that the union is over the speaker's right shoulder, and would appear to the audience to be on the left side of the flag. This applies whether the flag is hanging horizontally or vertically.
- There are other specialized flags in the Navy:
  —Signal flags are (still) used to communicate ceremonially and tactically.
  —A *commission pennant* flies night and day on commissioned Navy ships.
  —Admirals have their own flags with the same number of stars as they wear as rank; the term "flag officer" is derived from this.
  —Non-admiral officers in command of groups of more than one ship or aircraft rate either a *broad command* pennant or a *burgee command* pennant.

—Four different *absentee* pennants are flown to indicate the absence of the captain or embarked officials from their ship.

—Special flags are flown by a command to indicate that it has received a *unit award* (such as the Meritorious Unit Commendation).

—Each service has its own flag and, when displayed together, they must be placed in the following order: Army, Marine Corps, Navy, Air Force, Coast Guard. (The same rules apply as described above for the national ensign; i.e., the Army flag would be placed to the right of the Marine Corps flag, etc.)

—Battle streamers are attached to the top of service flags to represent wars, battles, campaigns, and so on.

- The word "yes" is used affirmatively as in civilian life, but "aye, aye" is used when a Sailor needs to communicate that he or she has heard, understood, and will carry out an order. "Sir" or "ma'am" is appended as appropriate.

- Seniors close memos, e-mails, and so on with "respectfully," whereas juniors use "very respectfully" when writing to seniors.

- "Attention on deck" is often called when a senior enters a room; all should stand until invited to sit.

- "Gangway" is often called to clear a path for a senior officer. "Coming through" is appropriate to clear a path for others; "make a hole" is not correct, but you will often hear it used. In all cases, you should move out of the way.

- In boats and vehicles, the general rule is *seniors in last and out first.*

- In boats, seniors usually sit aft of juniors.

- Certain words (like "coxswain" and "forecastle") are pronounced in strange ways.

- There are some odd spellings in the Navy as well (like "Alfa" and "Juliett").

- "Anchor's Aweigh" (*not* "Anchors Away") and "The Navy Hymn" (sometimes called "Eternal Father") are special songs that you will hear frequently in the Navy.

- Frocking occurs when someone is allowed to wear the next senior rank (without added pay).

- When a ship returns from a long deployment (and sometimes to honor a special dignitary), the crew lines up (evenly spaced) along the visible decks and levels; this custom is called "manning the rail."

- Ships are *dressed* or *full-dressed* with extra flags on certain holidays and for special occasions (like a visit to a port other than the ship's home port).
- When warships pass close aboard while underway, they will often show respect by executing *passing honors*; the junior ship will initiate and all hands visible on deck will stand at attention and salute on command.
- *Tending the side* is a means of honoring important visitors and is accomplished by having a certain number (depending upon the person's rank or position) of Sailors (traditionally called "side boys") lined up evenly to create a human passageway through which the visitor is *piped aboard* (saluted while a boatswain's whistle is blown).
- A number of "milestones" are traditionally recognized in the Navy (usually with a degree of humor and celebration):

  —*Crossing the Line*—Those who cross the equator on a ship are no longer *pollywogs* and are forever more *shellbacks*.
  —*Golden Dragon*—One who crosses the 180th meridian (international date line).
  —*Golden Shellback*—One who crosses the equator at the intersection of the 180th meridian.
  —*Emerald Shellback*—One who crosses the equator at the intersection of the Greenwich meridian (0ring longitude).
  —*Blue Nose*—One who crosses the Arctic Circle.
  —*Red Nose*—One who crosses the Antarctic Circle.
  —*Moss Back*—One who travels around Cape Horn.
  —*Order of the Ditch*—Recognizes a passage through the Panama Canal.
  —*Order of the Spanish Main*—Recognizes a cruise through the Caribbean Sea.
  —*Mystic Society of One Thousand and One Nights*—Recognizes a cruise through the Persian Gulf (also known as a "Persian Excursion").
  —*Realm of the Czars*—Recognizes a cruise through the Black Sea.

- Other recognized (through certificate and ceremony) achievements include:

  —*Deep Dive Diploma*—Commemorates first and last dives and other submarine occasions.
  —*Century Club*—Commemorates a naval aviator's hundredth carrier landing.
  —*200 (or 300) Mission Club*—Recognizes milestones in aviation missions.

—*Not So Ancient Order of the Hurriphooners*—Recognizes hurricane hunters who have flown through winds of one hundred miles per hour or more.

• Dubious "honors" include the *Goldfish Club* (which recognizes one who has ditched at sea), *Sea Squatters* (who have ditched and spent more than twenty-four hours in a life raft), and the *Caterpillar Club* (for those who have made an unscheduled parachute jump from a disabled aircraft).

CHAPTER SIX

Ships

The oldest element of sea power is the ship. From the rowed galleys of ancient Greece, to the sailed frigates of the Napoleonic era, to the nuclear-powered aircraft carriers of today, ships have always been the backbone of any navy.

Having some basic knowledge of the ships that give the Navy much of its formidable capability goes a long way toward understanding what the Navy is all about.

Countless volumes would be required to cover all the myriad details of Navy ships, but here we will provide the basics to get you started and allow you to function with a measure of understanding and credibility in the complex, technical world of the Navy.

This chapter deals with ships from an *external* view, explaining how to describe them and providing a basic understanding of their naval uses. For those who actually go aboard ships, another chapter describes ships with a more detailed *interior* view.

WHAT TO CALL SHIPS

One of the quickest ways to establish yourself as a naval novice is to refer to a ship as a boat. Unfortunately, there is no absolute way to define the difference, but the following guidance will work most of the time. In general, a boat is a watercraft that is small enough to be carried aboard a larger vessel, and that "larger vessel" is a ship. This is sometimes expressed this way: "A ship can carry a boat, but a boat can never carry a ship." Also it is helpful to remember that if a vessel has a permanent crew with a commanding officer assigned, it is more than likely a ship. If a vessel is only manned part of the time (when it is in use), it is probably a boat. Another distinction sometimes made is that a ship is designed to "navigate in deep waters," but there are some pitfalls with this, too. The best rule that works most of the time is that *if it is big, it is a ship*. Don't call a frigate or a destroyer or a cruiser a "boat."

Now for some confusion. Using the above guidance, submarines are technically ships. Yet they are traditionally referred to as boats. The original submarines were very small and manned only when in use, so "boat" was appropriate. But as they developed into larger vessels, and should rightfully have been called "ships," the original term stuck. There was an attempt by some submariners to change them over to "ships" when the large nuclear subs began to appear, but as with many things in the Navy, tradition trumped logic, and today, all submarines—even the giant "boomers" (fleet ballistic missile submarines)—are called boats.

Another exception is that sometimes personnel who are assigned to air wings embarked in aircraft carriers will refer to the carrier as "the boat." There is no official sanction for this, but it seems to be a kind of affectionate irreverence that they use to set themselves apart from their fellow Sailors who are assigned to the carrier as a part of its permanent crew. Make no mistake: an aircraft carrier is a ship—let others call it a boat if they must.

One term that causes some consternation in naval circles is the word "vessel." There are some cantankerous would-be purists who insist that a vessel is "something used to carry water, not to sail upon it." But the *Dictionary of Naval Terms* defines "vessel" as "every description of craft, ship, or other contrivance used as a means of transportation on water." Other dictionaries (including Webster's: "a watercraft bigger than a rowboat") confirm the acceptability of this term, and "The Official Inventory of U.S. Naval Ships and Service Craft" is officially known as the "Naval Vessel Register." So, the bottom line is that "vessel" is an acceptable term in most naval circles, but if you happen to work for one of the aforementioned "purists," I would not advise arguing with him or her about it.

By way of interest (and now getting into the real purist category), the term "ship" did not always have the generic meaning it now does. In the days of sail, a ship was a vessel with three masts carrying mostly square sails ("square rigged"). Over time, the term has become the encompassing one we hear today.

One last comment regarding ship references. Tradition has long mandated the use of feminine pronouns when referring to ships (as in, "She has a new sonar, making her a good ASW ship"). This practice may be a dying one

(Lloyds of London, the long-standing maritime British insurance company, no longer uses the feminine), and you will no longer be "keelhauled" for not using it, but you will still frequently encounter the practice, so you should be aware of it. Being a traditionalist (a.k.a. "dinosaur"), I use feminine ship references in this book. Whether to use them yourself or not is largely a function of the old adage, "Know your audience."

HOW SHIPS ARE EMPLOYED

The U.S. Navy operates hundreds of ships. Some of these are *active* ships, which means they have a full complement of active duty personnel (crew) and, unless they are temporarily undergoing heavy maintenance or repair, are fully capable of carrying out an assigned mission on short notice.

The Navy also keeps a number of vessels in *reserve* status, which means that they are fully functional ships but are only partially manned with active duty personnel. The rest of the crew is made up of reserve personnel, who only man the ships periodically for training and when called upon in national emergencies. For this reason, these ships are mission-capable and are an essential part of the Fleet but are usually less readily available than are ships in active status.

The Navy also operates a number of vessels for the Department of Defense under what is called the "Military Sealift Command" (MSC). These ships usually have only a very small contingent of Navy personnel on board, and the majority of the crews are civilians. MSC ships have a support role and are not used as front-line combatants. They are considered to be "in service" rather than "in commission." Many of these ships, such as vehicle cargo ships and transport oilers, serve the Army and Air Force as well as the Navy.

Other MSC ships perform special-duty projects, such as ocean-bottom laying and repairing of cables used for detecting enemy submarines. Surveying ships and oceanographic research ships explore the oceans and provide important hydrographic information for navigational and other purposes.

Of special interest is a group of various MSC ships that make up the Naval Fleet Auxiliary Force (NFAF). These ships are the lifeline to U.S. Navy ships at sea, providing fuel, food, ammunition, spare parts, and other supplies to the operating forces, alleviating the need for them to constantly return to port for supplies. As with other MSC ships, they have civilian officers and crews, but they operate under Navy orders and have a contingent of Navy personnel aboard, performing visual and radio communications and otherwise assisting the ship's civilian master and crew in coordinating operations with other naval units.

WAYS TO DESCRIBE SHIPS

You don't have to see very many Navy ships before you realize that there are many different types. There are a number of ways to describe them, some generic and some more specific, dealing with such things as size and capabilities.

Dimensions

A ship's displacement is for all intents and purposes (except to naval architects and other purists) her weight.

A ship's length is called just that ("length"), but her width is called beam.

Draft is how deep into the water a ship's hull reaches and is generally expressed in feet; obviously it would not be a good idea for a ship with a draft of twenty-five feet to venture into waters with a depth of twenty feet.

Speed

Ships have different speed capabilities, depending upon their mission. When describing a ship's speed, it is given in knots, which are slightly more than miles-per-hour: one knot = 1.152 miles per hour (or 1.85 kilometers per hour).

Propulsion

Various kinds of propulsion are used in today's ships, including steam, nuclear, gas turbine, and diesel.

Steam

The primary method of marine propulsion for more than a century, early steam plants burned wood or coal in boiler furnaces to heat water until it became steam. Most modern steam plants use oil as the heat source. The steam is then converted into usable energy by turbines that drive the propeller shafts.

With near-infinite quantities of water surrounding ships at sea, it would seem that a water supply for steam plants is not a problem. But salt water does not work well in steam systems, so ships must use fresh water and obviously cannot carry unlimited amounts. To preserve the precious fresh water, shipboard steam plants operate as a closed cycle—meaning that the water and steam are theoretically contained in the system and not allowed to escape. Steam is retrieved from the turbines and returned to condensers that convert the spent steam back into fresh water, which is then returned to the boilers to be reheated into energy-filled steam again. This "steam cycle," as it is called, is repeated over and over to propel a ship through the water. Even though this is a closed cycle, a certain amount of the fresh water is used up, so that a continuous supply of feedwater is required for sustained operations. Shipboard distilling plants create the feedwater by converting sea (salt) water into fresh.

The fuel needed is carried in fuel tanks on board—much as an automobile has a gas tank—and must be periodically replenished (either in port or from oilers at sea).

Nuclear Power

Although using very sophisticated technology, shipboard nuclear power plants are actually just a variation of steam propulsion. Instead of using oil-fired boilers, nuclear-powered ships have reactors that produce the heat to convert fresh water to steam.

Nuclear power gives a ship the advantage of great endurance at high speed. Instead of refueling every few thousand miles like an oil-burning ship, a nuclear-powered ship can operate for years on one reactor core, so it can steam almost indefinitely, limited only by its need to replenish food and spare parts (and ammunition in wartime).

Another favorable feature of nuclear power is that, unlike conventional oil-fired systems, the generation of nuclear power does not require oxygen. This makes it particularly useful as a means of submarine propulsion. Nuclear-powered submarines can operate completely submerged for extended periods of time.

Gas Turbines

This modern form of propulsion uses jet engines that are very much like the ones used in aircraft but have been adapted for use on ships. The burning fuel spins turbines in the engines that convert the energy created into usable power that turns the ship's propellers.

Although some of the principles are the same, some of the primary differences between these propulsion plants and those that use steam are that the gas turbines combine the functions of the boiler and the turbines into one element, and gas turbines have no need of feedwater. This means that they are smaller, more efficient, and easier to maintain. They are also much more quickly "brought on the line" (turned on). A steam-powered vessel requires hours to prepare to get underway, whereas gas turbine-powered ships can be ready in minutes.

The obvious advantages of gas-turbine technology have caused the U.S. Navy to build more and more of these ships. Whereas steam was once the main means of naval propulsion, today there are more gas-turbine ships in the Navy than any other kind of propulsion.

Diesel Engines

Frequently used in ships that need less horsepower, diesels are lighter, take up less space, and are more efficient than steam turbines. Marine diesels are basically larger versions of the engines used in trucks, buses, and some automobiles, except that their power is put to propellers rather than wheels. The diesel engine can be coupled directly to the propeller shaft through reduction gears and perhaps a clutch; or it can drive a generator that produces electrical current for the main drive.

Diesel engines are preferred over gasoline engines because they are more efficient and because diesel fuel is not as volatile as gasoline. An additional problem with gasoline is that its fumes are heavier than air and tend to collect in low places in a ship, making them very dangerous.

Sails

Utilizing wind to propel ships through the water, sails were once the primary means of propulsion for ships. Although naval vessels no longer use them, with a nod toward tradition, Navy ships are often said to "sail" from one place to another (as in, "The fleet sailed for the Middle East").

Another generic term that is still often used in the same context (despite the proliferation of gas turbine ships) is "steam" (as in, "The destroyer steamed into the Persian Gulf").

So far, no one has used "turbined" in this context (and probably never will).

Missions

Ships' primary missions are often described by acronyms. If a ship is capable of engaging aircraft or incoming missiles, she is said to be an AAW (antiair warfare) ship. Her ability to engage other ships is described as ASUW (anti-surface warfare) and opposing enemy submarines is known as ASW (antisubmarine warfare). The term USW (undersea warfare) is a more recent term that includes mine warfare (MIW) as well as ASW. Strike warfare refers to the ability to attack land targets.

A number of things determine a ship's mission capabilities, including the weapons she is armed with and the sensors (radar, sonar, etc.) she carries. These things will be addressed in subsequent chapters.

HOW THE NAVY IDENTIFIES SHIPS

You will often see U.S. Navy ships identified by a combination of letters and numbers such as

USS Enterprise (CVN-65)

Most Navy ships have both a *name* (such as "*Enterprise*") and what we call a "ship's designation" (such as "CVN-65") to identify them. Though the name is a convenient and traditional means of identification, there have been many Navy ships bearing the same names throughout history, so the ship's designation—which is unique to each individual ship—is the only way to positively identify a specific naval vessel. The ship's designation tells what *type* (in this case, "CVN" identifies her as a nuclear-powered aircraft carrier) and assigns a specific *hull number* (in this case, "65") to the vessel.

Ships are also grouped into *classes* to identify those with identical, or nearly identical, characteristics.

Name

The name is unique to a ship in that there can only be one Navy ship in commission at a time with a given name. But, as already mentioned, there may have been other ships with the same name in the past—in fact, it is fairly common practice in the Navy for ships to carry the name of an earlier ship who served with honor. For example, there have been eight U.S. Navy ships named *Enterprise*.*

The name of a Navy ship in commission (active or reserve) is preceded by the letters "USS," which stands for "United States Ship." By convention, the name (but not the "USS") is usually written in *italics*, or both are written in all capital letters (such as USS ENTERPRISE). The "USS" designation also applies to submarines, even though, as earlier noted, they are referred to as "boats."

MSC ships (see chapter 2) are somewhat different. Because they are considered to be "in service" rather than "in commission," names of MSC ships are preceded by the letters "USNS" (for United States Naval Ship) instead of "USS."

U.S. Coast Guard operates ships carrying the prefix USCGC, for "United States Coast Guard Cutter."

Other navies of the world use similar systems. In the Royal Navy, vessels carry the prefix "HMS," which stands for "Her Majesty's Ship" (or "His Majesty's Ship" if there is a reigning king instead of a queen). The navies of other nations are similarly identified.

The Secretary of the Navy is the person responsible for naming U.S. Navy ships, although he or she gets plenty of help (from politicians, historians, admirals, families of famous people, and so on, who all have their own ideas of what the next ship should be named). There are some "rules" (actually conventions) to ship naming, although these are sometimes broken and have changed over the years. At one time, battleships were named for states and cruisers for cities, but today submarines are mostly named for states with some named for cities (*Henry M. Jackson*, *Jimmy Carter*, and *Seawolf* being notable exceptions). Aircraft carriers are now mostly named for presidents, though there is one admiral (*Nimitz*) and two congressmen (*John C. Stennis* and *Carl Vinson*), with *Enterprise* and *Kitty Hawk* being exceptions. Cruisers are named for important battles in U.S. history. Destroyers are named for people—Sailors or Marines who have served or sacrificed exceptionally (*Winston S. Churchill* being an exception). Amphibious ships are a mix of battles

---

*This count of course does not include the starship *Enterprise* of *Star Trek* fame, but the creator of the hit television and movie series, Gene Roddenberry, recognized the long tradition of passing on ship names and carried it on in his futuristic vision.

(*Tarawa, Saipan*, etc.), cities (*Anchorage, Nashville*, etc.), famous American landmarks (*Rushmore, Hermitage*, etc.), and former famous ships (*Kearsarge, Bon Homme Richard*, etc.). Other types of ships follow other conventions as well (minehunters are named for birds, for example).

Sailors traditionally add nicknames to their seagoing homes. Among aircraft carriers, for instance, the USS *Enterprise* is known informally as the "Big E," USS *Theodore Roosevelt* is "TR," and *Dwight D. Eisenhower* is "Ike." Some of these nicknames are more "colorful" and it is probably better not to record them here.

Designation

Though a ship's name gives her some identity, her *designation*—which consists of a combination of letters and numbers—is a unique identification that tells you two additional things about the ship: her type and her place in the construction sequence. The USS *Theodore Roosevelt*, for instance, has the designation CVN-71. CVN is her type classification, "CV" standing for aircraft carrier and "N" meaning nuclear propulsion. The number 71 indicates that she is the seventy-first aircraft carrier authorized for construction. The term "hull number" actually refers only to the number part of the ship's designation, but you will commonly hear it used instead of "ship's designation," referring to the letter and number combination (as in, "USS *Enterprise*'s hull number is CVN-65"). Ships' hull numbers (numerals only) are frequently painted on their bows and near the stern. Aircraft carriers have their hull numbers painted on the forward part of the flight deck and on the "island" (superstructure).

Since 1920, the Navy has used letter symbols to identify the types of ships, boats, and service craft. This is called "type classification" and is used as part of the ship's designation. Some of the more common type classifications are listed below. Keep in mind that some of these type classifications may not be currently be in use, but they are listed because you may come across them historically, or they may be reactivated at some later date.

AD	destroyer tender
AE	ammunition ship
AFS	combat store ship
AGF	miscellaneous command ship
AH	hospital ship
AKA	attack cargo ship

AO	oiler
AOE	fast combat-support ship
AOG	gasoline tanker
AOR	replenishment oiler
APA	attack transport ship
APB	self-propelled barracks ship
APL	barracks craft (non-self-propelled)
AR	repair ship
ARS	salvage ship
AS	submarine tender
ASR	submarine rescue ship
ATF	fleet ocean tug
AVT	training aircraft carrier
BB	battleship
CA	heavy cruiser
CC	command cruiser
CG	guided-missile cruiser
CGN	guided-missile cruiser (nuclear propulsion)
CL	light cruiser
CLAA	antiaircraft cruiser
CLGN	guided-missile light cruiser (nuclear propulsion)
CV	multipurpose aircraft carrier
CAG	guided-missile heavy cruiser
CVA	attack aircraft carrier
CVL	light aircraft carrier
CVN	multipurpose aircraft carrier (nuclear propulsion)
CVT	training aircraft carrier
CVS	antisubmarine warfare aircraft carrier
DD	destroyer
DDG	guided-missile destroyer
DDR	radar picket destroyer
DD(X)	prototype destroyer
DE	destroyer escort (also ocean escort)
DEG	Guided-missile ocean escort
DER	radar picket destroyer escort
DL	destroyer leader (once also called "frigate")
DLG	Guided-missile destroyer leader (once also called "guided-missile frigate")

DLGN	nuclear-powered guided-missile destroyer leader (also "frigate")
DM	destroyer minelayer
DSRV	deep-submergence rescue vehicle
FF	frigate
FFG	guided-missile frigate
HSV	high-speed vessel
IX	unclassified miscellaneous
LCAC	landing craft, air cushioned
LCC	amphibious command ship
LCH	landing craft, heavy
LCIL	landing craft, infantry, large
LCM	landing craft, mechanized
LCPL	landing craft, personnel, large
LCS	littoral combat ship
LCU	landing craft, utility
LCVP	landing craft, vehicle and personnel
LHA	amphibious assault ship (general purpose)
LHD	amphibious assault ship (multipurpose)
LKA	amphibious cargo ship
LPD	amphibious transport dock
LPH	amphibious assault ship (helicopter)
LSD	dock landing ship
LSIL	landing ship, infantry, large
LSM	landing ship, medium
LSSC	light SEAL support craft
LSSL	landing ship, support, large
LST	tank-landing ship
MCM	mine-countermeasures ship
MCS	mine-countermeasures support ship
MHC	coastal minehunter
MSC	coastal minesweeper
MSO	ocean-going minesweeper
NR	submersible research vessel
PB	patrol boat
PBR	river patrol boat
PC	coastal patrol craft
PCF	fast patrol craft ("swift boat")

PG	patrol gunboat
PHM	patrol hydrofoil missile
PT	patrol torpedo boat
SS	submarine
SSG	guided-missile submarine
SSBN	fleet ballistic-missile submarine (nuclear powered)
SSGN	guided-missile submarine (nuclear powered)
SSN	submarine, attack (nuclear powered)
YD	floating crane
YP	yard patrol
YTB	large harbor tug
YTL	small harbor tug
YTM	medium harbor tug

Ships of the Military Sealift Command (MSC) are distinguished from other Navy ships by having a "T" before their letter designations. Below are some examples of MSC ship types.

T-ACS	crane ship
T-AE	ammunition ship
T-AFS	combat stores ship
T-AGM	missile range instrumentation ship
T-AGOS	ocean surveillance ship
T-AGS	oceanographic survey
T-AH	hospital ship
T-AK	maritime pre-positioning ship
T-AKR	vehicle cargo ship
T-AO	oiler
T-AOT	tanker
T-AP	troop ship
T-ARC	cable repair
T-ATF	fleet ocean tug
T-AVB	aviation logistic ship

Coast Guard cutters also use this type of classification and are distinguished by having "W" as the first letter in their designation. Examples of Coast Guard ship types are:

WHEC   high endurance cutter
WMEC   medium endurance cutter
WAGB   ice breaker
WLB   seagoing buoy tender
WLM   coastal buoy tender
WPB   patrol boat

Class

Within a type classification of vessels there are classes. Ships belonging to a particular class are built from the same plans and are very much alike, essentially identical except for the different hull number painted on their bows and a different assigned name. In reality, one can always find some minor differences and occasionally, individual ships within a class may be significantly altered. The first ship built to a specific design determines the name of the class. For example, after World War II the United States redesigned its aircraft carriers to accommodate the newly invented jet aircraft then entering the Fleet. The first of these new aircraft carriers to be built was commissioned as USS *Forrestal* (CV-59). She was the fifty-ninth aircraft carrier, but the first of this new class. Satisfied with these new ships, the Navy built three more—USS *Saratoga* (CV-60), USS *Ranger* (CV-61), and USS *Independence* (CV-62)—all of which are referred to as *Forrestal*-class carriers.

Later, some major improvements were deemed necessary, so the Navy redesigned its aircraft carriers significantly enough that they were considered a new class of carrier. The first of these new and different carriers was named USS *Kitty Hawk* (CV-63), so the next ship built after her, USS *Constellation* (CV-64), was considered a *Kitty Hawk*—class aircraft carrier. And so the process goes.

Sometimes you will hear a class identified by the hull number of the first ship. The *Oliver Hazard Perry* class of guided-missile frigates is often referred to as the "FFG-7-class," and the *Los Angeles* class of nuclear attack submarines are often called the "688s."

Ship Types and Their Missions

The many different types of vessels in the Navy have specific functions or missions. Some exist primarily to engage in combat with enemy forces (other vessels, aircraft, or land targets) and are generally referred to as *combatants*. Others, known as *auxiliaries*, exist to deliver the supplies (fuel, ammunition, food, and repair parts) needed by the operating forces, to provide maintenance and repair services, to conduct salvage operations, and to provide a host of other support functions. Still others, known as *amphibious* vessels, are designed to take troops where they are needed and get them ashore, and *mine-warfare* vessels locate and destroy underwater mines.

The major combatants can be divided into three groups: *aircraft carriers*, *submarines*, and *surface combatants*. The latter include (from largest to smallest) *cruisers, destroyers, frigates*, and (in development) *littoral combat ships* (LCS).

The largest surface combatants ever built were the *battleships*. They played significant roles in naval combat for much of the twentieth century but have not survived into the twenty-first (except as museum ships). There is a tendency among the uninformed (television reporters, for example) to refer to any combatant as a "battleship"; such references make the informed cringe, so it is to be avoided.

Aircraft Carriers

These gigantic ships have been described as the world's largest combatant ships and the world's smallest airfields. Their displacement is between eighty and one hundred thousand tons (depending upon the class), they are more than a thousand feet long, and they have beams exceeding 250 feet. They carry about eighty-five aircraft, and the number of personnel required to operate the ship and its aircraft is more than five thousand. Some of the U.S. Navy's carriers are driven by oil-fired boilers and are designated CV, whereas others are nuclear powered and so designated CVN; all have four screws (propellers) capable of driving them at speeds greater than thirty knots (Photo 6.1).

Aircraft carriers carry an assortment of aircraft capable of performing a wide variety of missions, including air support to troops ashore, bombardment missions, antisubmarine operations, rescue missions, reconnaissance, and antiair warfare. They are capable of staying at sea for long periods of time, making them potent weapons in a wide variety of scenarios.

Though it is not entirely incorrect to refer to the Navy's helicopter-carrying amphibious ships (LHAs and LHDs) as "aircraft carriers," that term is normally reserved for the larger predominantly jet-carrying ships (CVs and CVNs).

Aircraft carriers are frequently referred to as just "carriers," sometimes by the nickname "flattop," and sometimes (irreverently) as "bird farms."

Submarines

Though in many ways submarines are ships, they are different enough to warrant a separate treatment. As explained earlier, despite their size and complex capabilities, they are traditionally called "boats."

The U.S. Navy has three major types of submarine—SSN, SSBN, and SSGN—all of which are nuclear powered. The *attack submarine*, designated SSN, has the primary mission of other submarines and surface ships, but they are also assigned secondary missions, which may include surveillance and reconnaissance, direct task-force support, landing-force support, land attack, mine laying, and rescue. The SSN's principal weapons are high-speed, wire-guided torpedoes and cruise missiles for use against surface and land targets.

The *fleet ballistic-missile submarines* (SSBN) have a strategic mission, in that they are meant to deter or participate in a nuclear-missile exchange. They are the "sea leg" of the U.S. nuclear strike triad (the other two components being Air Force strategic bombers and land-based ICBM systems). Their highly sophisticated, very potent ICBMs have multiple nuclear warheads and are capable of hitting large targets many thousands of miles away and causing tremendous destruction. You will often hear them referred to by the nickname "boomer boat" or just "boomer" (Photo 6.2).

SSBNs must remain submerged for long periods of time, virtually out of contact with the rest of the world, serving primarily as a deterrent but essentially waiting to carry out a mission that could be devastating to much of the world. This is a stressful environment for the crews and, to alleviate some of that stress, SSBNs are operated during alternate periods by two separate crews. One is called the blue crew and the other the gold crew. On return from an extended patrol, one crew relieves the other, and the ship returns to patrol following a brief period alongside her tender or in port. The relieved crew enters a month-long period of rest, recreation, and leave, followed by two months of training. This system allows each crew time ashore, while keeping these important ships cruising on deep patrol except for very brief periods.

The third major type of submarine currently in the U.S. arsenal is the *guided-missile submarine* (SSGN). Converted *Ohio*-class SSBNs, these submarines are armed with *Tomahawk* tactical missiles for land attack missions and are designed for a variety of specialized missions, including the ability to transport, insert, and support SEALS or other special operations forces for prolonged periods. Secondary missions are the traditional attack submarine

Photo 6.1   Nuclear-powered aircraft carrier (CVN) (PH3 Greg Welch)

Photo 6.2   Fleet ballistic-missile submarine (SSBN) (PH1 Michael J. Rinaldi)

Photo 6.3   Cruiser (CG) (U.S. Navy, Christopher Mobley)

Photo 6.4   Guided-missile destroyer (DDG) (U.S. Navy, Frederick McCahan

Photo 6.5 Frigate (FFG) (U.S. Naval Institute)

Photo 6.6 Landing ship dock (LSD) (U.S. Navy, Mahlon K. Miller)

Photo 6.7 Amphibious transport dock (LPD) (Northrop Grumman Corporation)

Photo 6.8 Multipurpose assault ship (LHD) (U.S. Navy)

missions of intelligence, surveillance, and reconnaissance (ISR), battle space preparation, and sea control.

Cruisers

These powerful ships are extremely capable in AAW, ASW, and ASUW missions. They are equipped with missiles that can knock out incoming raids from enemy aircraft or missile attacks. With other specially designed missiles, they are able to hit land or sea targets at substantial distances.

Currently, the Navy's cruisers are all *Ticonderoga*-class (designated CG) ships, which are powered by four gas turbines driving twin screws at speeds greater than thirty knots and are equipped with the very sophisticated Aegis combat system (Photo 6.3). This integrated combat system is highly automated, exceptionally fast, and capable of conducting antiair, antisurface, and antisubmarine warfare simultaneously. These ships fire a variety of missiles (Tomahawk and Standard) as well as torpedoes and ASROC (antisubmarine rocket) and have two MK 45 5-inch guns. They also are equipped with two Phalanx close-in weapons systems (a rapid-firing gun that is used to knock down an incoming missile if other systems have failed to bring it down).

Destroyers

Destroyers have always performed a wide range of missions. They can serve as part of a screen unit in a carrier task group, protecting it from various forms of attack. They can detect and engage enemy submarines, aircraft, missiles, and surface ships. In an amphibious assault, a destroyer's weapons help protect against enemy forces at sea and ashore. In short, destroyers have a well-deserved reputation of being the "workhorses" of the Fleet.

Earlier classes of destroyer were rather small—some displacing as little as four hundred tons—but today's *Arleigh Burke*-class destroyers displace as much as ninety-two hundred tons (Photo 6.4). They are a little more than five hundred feet in length with a beam of nearly sixty feet. These powerful ships are powered by four gas turbine engines that drive two screws and make the ship capable of speeds greater than thirty knots. Like the *Ticonderoga*-class cruisers, they are equipped with the Aegis combat system, making them the most potent class of destroyer ever built. At one time, the differences between cruisers and destroyers were significant. Today, the differences are not so obvious.

Earlier versions of the *Arleigh Burke* class have a flight deck but no hangar whereas the newer versions do have a hangar. They are capable of firing a variety of missiles (Harpoon, Tomahawk, Standard, and Evolved Sea Sparrow) against ships, land targets, aircraft, and enemy missiles. They can also fire torpedoes and the ASROC to destroy enemy submarines, and they have the powerful Mk 45 gun system. They also are equipped with the Phalanx close-in weapons system.

A new type of destroyer called the DD(X) and named the *Zumwalt* class is a highly capable platform that includes sophisticated "stealth" (low susceptibility to radar detection) technology, efficient electric drive systems, low manning requirements, a very long-range gun system, and other impressive innovations.

Destroyers are sometimes called "tin-cans," a traditional nickname that was much more appropriate to earlier classes than to the highly sophisticated marvels of today.

Frigates

Originally a type of sailing warship, modern frigates first appeared in the U.S. Navy during World War II as destroyer-like ships (but smaller) and were known then as destroyer escorts (DE). Some were called "ocean escorts" for a time, then in 1975 they were re-designated as frigates (FF) and then as FFGs once guided missiles were added (Photo 6.5). Frigates fulfill a protection of shipping (POS) mission as antisubmarine warfare (ASW) combatants for amphibious expeditionary forces, underway replenishment groups, and merchant convoys. They are a good general purpose small vessel that will be phased out when the new LCS ships come on line in sufficient quantity. The guided-missile frigates (FFG) also once offered a limited antiair warfare (AAW) capability but as a cost-saving measure, the Navy has removed the missiles they once had, so now they now are often referred to as the "FF-not-so-G"s.

A number of classes have been built over the years, but today the *Oliver Hazard Perry*—class frigates are the only ones in commission in the U.S. Navy. These ships carry crews of a little more than two hundred and may be viewed as scaled-down destroyers. Despite their limited capabilities compared to the larger destroyers and cruisers, the current class of frigates is a tough platform, capable of withstanding considerable damage, as aptly demonstrated when USS *Samuel B. Roberts* struck a mine and USS *Stark* was hit by two Exocet cruise missiles in the Persian Gulf. In both cases the ships survived, were repaired, and returned to the Fleet.

Like the current cruisers and destroyers, they are powered by gas turbine engines but have only one screw (propeller). They are about 450 feet long and have a beam (width) of forty-five feet. They displace (weigh) forty-one hundred tons and have a top speed of just under thirty knots. They can carry one or two helicopters, have two triple-tubed torpedo launchers, can fire antiair or surface-to-surface missiles, and have one 76-mm gun and a Phalanx close-in weapons system.

Designed as cost efficient surface combatants, the FFGs lack the multi-mission capability necessary for modern surface combatants faced with multiple, high-technology threats and they offer limited capacity for growth, so it is unlikely the Navy will build any more of them.

One point of possible confusion that you might need to be aware of is that shortly after World War II, the Navy built some "super destroyers" that were bigger and more capable than normal destroyers. These were designated "DL" (and "DLG") and were called "destroyer leaders." Unfortunately, they were also called "frigates." This was the exact opposite of what the rest of the world was doing at the time—calling ships *smaller* than normal destroyers "frigates"—so it caused some confusion, particularly when ships of different navies began operating together during the Cold War as part of NATO and other alliances. Eventually, these destroyer leaders were re-designated as cruisers and the name "frigate" was given to the smaller destroyer escorts. Bottom line: if you see the term "frigate" used in reference to a U.S. Navy ship prior to 1975, it is referring to a vessel *bigger* than a destroyer, and after 1975, it is referring to a ship *smaller* than a destroyer.

Littoral Combat Ships

Described by the Secretary of the Navy as "fast and capable ships [that] will . . . provide us with an ability to operate in the littoral areas of the world where the enemies of freedom seek to operate and hide," the basic idea behind this new type of ship is to allow the U.S. Navy to operate in closer to shore than it has traditionally. Although there are times in the Navy's past when it has operated in these areas (Civil War, Vietnam, etc.), the U.S. Navy has been primarily a "blue-water" (meaning "deep ocean") fleet for the majority of its existence. With the rise of terrorism, the need for projecting power ashore as part of joint operations, and the current lack of a blue-water competitor, the Navy now focuses more attention on the coastal regions of the world. "Littoral" is used rather than "coastal" because it is meant to convey a broader area. "Coastal" more precisely defines the *edge* where land and sea meet, whereas "littoral" is meant to include more of the adjacent waters and the land area where power must be projected in the accomplishment of various missions in the modern environment. The older term "brown-water operations" (generally describing coastal and riverine operations) comes close but is too "wet," not embracing the adjacent land areas as an integral component.

The littoral combat ship (LCS) is therefore a new breed of vessel. Designed to be fast and maneuverable as well as being stealthy (difficult to detect on radar) and having a shallow draft, these vessels incorporate modularity to make them flexible. This simply means that they can be rigged with different "packages" of electronics, weapons, and so on to tailor them to specific missions and that these packages can be changed relatively easily as different missions need to be carried out. These packages include deployable manned and unmanned vehicles (boats and aircraft) that extend the reach of these vessels in the littoral. Among the different missions these vessels are able to carry out are Special Forces insertion and support, reconnaissance, maritime interdiction (search and seizure of other vessels), mine countermeasures, and antiterrorism force protection. These capabilities permit the Navy to penetrate areas that have long been off limits or very high risk for the larger (very expensive) combatants (cruisers, destroyers, etc.).

Auxiliaries

Besides the combatants, the Navy has a large number of auxiliaries that provide support services to the Fleet.

Underway replenishment (UNREP) ships allow combatants to remain at sea for long periods of time without having to return to port by bringing fuel, provisions, repair parts, and ammunition to the ships and transferring them at sea. The U.S. Navy is highly proficient at underway replenishment techniques, using special cargo-handling gear to make transfers from one ship to another while the two are steaming abreast or, in some rare cases, astern. Vertical replenishment (VERTREP) is a form of UNREP in which cargo-carrying helicopters are used to transfer goods from one ship to another. In the past, underway replenishment ships were part of the operating fleet, but today most of the UNREP delivery capability of the Navy is carried out by MSC ships.

Submarine tenders (AS) are full of maintenance and repair shops and are manned by technicians with a wide variety of skills so that vessels coming alongside can receive rather extensive repairs or have major maintenance performed on them.

Salvage vessels and rescue vessels provide rapid firefighting, dewatering, battle-damage repair, and towing assistance to save ships that have been in battle or victims of some other disaster from further loss or damage. Some have specialized equipment and are manned by salvage divers so that they can also perform submarine rescue and salvage operations underwater.

Also among the Navy's waterborne resources is a large and varied group of service craft. Some are huge vessels like the large auxiliary floating drydocks that can take very large vessels aboard and raise them out of the water for repairs. Barracks craft accommodate crews when their ships are being over-hauled or repaired. Lighters are barges used to store and transport materials and to house pier-side repair shops. Some gasoline barges, fuel-oil barges, and water barges are self-propelled; those that are not depend on tugs. Floating cranes and wrecking derricks are towed from place to place as needed. Diving tenders support diving operations. Ferryboats or launches—which carry people, automobiles, and equipment—are usually located at Navy bases where facilities are spread out over large distances. Best known of the service craft are the harbor tugs, large and small, that aid ships in docking and undocking, provide firefighting services when needed, perform rescues, and haul lighters from place to place.

Amphibious Warfare Ships

Often referred to as "amphibs" or "gators," these ships work mainly where sea and land meet, and where assault landings are carried out by Navy–Marine Corps teams. Such operations call for a variety of types of ships. Many are transports of varied designs, used to sealift Marines and their equipment from bases to landing beaches. The differences lie in ship design and the way troops and their gear are moved from ship to shore, which can be done by means of landing craft, helicopters, or tracked amphibious vehicles.

They include dock landing ships (LSD) (Photo 6.6) and amphibious transport docks (LPD) (Photo 6.7) that deliver troops and equipment primarily by landing craft, as well as general purpose assault ships (LHA) and multipurpose assault ships (LHD) (Photo 6.8) that have large flight decks for massive helicopter operations and can also accommodate vertical/short take-off and landing (V/STOL) jet aircraft to provide significant strike and support capability.

Mine-Warfare Ships

Two types of mine-warfare vessel are currently in use in the U.S. Navy: the *Osprey*-class minehunters with the designation MHC and the *Avenger*-class mine-countermeasures ships with the designation MCM. The MHCs specialize in detecting and locating today's highly sophisticated mines, and the MCMs are tasked with removing or destroying them. Specially configured helicopters also play a very large role in mine-warfare operations in the U.S. Navy. Specially designed ships used to be used to lay mines in enemy waters, but this

QUICKREFS

function, if employed, would be carried out primarily by aircraft or submarines today.

- The difference between ships and boats is not absolutely definable, but the following guidance will cover most instances:
  —Ships are bigger than boats.
  —Boats can often be carried on ships; ships are not carried on boats.
  —Ships usually have a permanent crew with a commanding officer, whereas boats usually have crews on board part of the time.
  —Submarines are called boats even though they meet the criteria for ships.
  —There are three kinds of Navy ships:
  —Active ships with full crews and ready on short notice to carry out assigned missions.
  —Reserve ships that are fully functional but only partially manned with active duty personnel with the rest of the crew consisting of reserve personnel.
  —MSC ships, which perform support functions and are primarily manned by civilians.
- Ships can be described in a number of ways:
  —by *displacement*, which is essentially a ship's weight;
  —by *length* and *beam* (width);
  —by *draft*, which is the depth a ship's hull reaches into the water;
  —by speed, which is expressed in *knots* (one knot = 1.152 miles per hour); and
  —by propulsion (of which there are four main types: *steam, nuclear, gas turbine,* and *diesel*).
- Ships are often said to *sail* (or *steam*) from place to place, no matter what their form of propulsion.

- Common ship missions are AAW (antiair warfare), SUW (surface warfare), ASW (antisubmarine warfare), USW (undersea warfare), and MIW (mine warfare).

- Most Navy ships, like USS *Enterprise* (CVN-65), have a *name* (*Enterprise* in this example) and a *designation*, consisting of a type (CVN in this case, symbolizing that she is a nuclear-powered aircraft carrier) and a *hull number* (65, indicating she is the sixty-fifth ship of that type).

- USS stands for "United States Ship" and indicates a ship is *in commission*. USNS stands for "United States Naval Ship" and indicates a ship is *in service* as part of the MSC.

- There have been many different designations used, but the ones you are most likely to encounter today are:

CV, CVN*	multipurpose aircraft carrier
SSN	attack submarine
SSBN	fleet ballistic-missile submarine
SSGN	guided-missile submarine
CG**	cruiser
DDG	guided-missile destroyer
FFG	frigate
LCS	littoral combat ship
LHA	amphibious assault ship (general purpose)
LHD	amphibious assault ship (multipurpose)
LPD	amphibious transport dock
MCM	mine-countermeasures ship

- Ships built to the same basic plans and nearly identical are said to be in the same "class."

- Besides the highly specialized aircraft carriers and submarines, today's Fleet consists of surface combatants (cruisers, destroyers, frigates, and littoral combat ships), auxiliaries (support ships), amphibious ships (used to put troops ashore and support them), and mine-warfare ships (which detect and neutralize enemy mines).

---

*The "N" appearing at the end of some of these designations indicates that the ship is nuclear powered.

**A "G" in the designation indicates that the ship is capable of firing guided missiles.

CHAPTER SEVEN

Shipboard

Many Navy civilian workers have occasion to go aboard ship or deal closely with them in some way. But even if you do not actually go aboard or work as a naval architect there is another reason that you might want to learn something about shipboard customs and terminology. In many Navy offices, agencies, and so on—though located many miles from any significant bodies of water and looking for all intents and purposes like any other building—you will hear the floors referred to as "decks," the stairwells as "ladders," and other such strange utterances. This stems from the presence of Sailors in these edifices and their need to remain connected in some way to their profession.

The question becomes, should civilians who may never have even seen a ship also use these terms. The answer is decidedly "yes." To do otherwise is unnecessarily alienating. Players on the same team should not use two different languages to describe their surroundings. If you worked in an art museum you would not do yourself any favors by referring to an oil portrait as a "picture."

The previous chapter—"Ships"—introduced you to ships so that you might be more comfortable with them when you encountered them on paper or in discussions. This chapter will give you a more intimate—internal—view of them for the reasons stated above.

We will begin with basic terminology and end with a guide for those who will have occasion to actually go aboard a ship.

SHIPBOARD TERMINOLOGY

No doubt you already have some knowledge of shipboard terminology even if you have never set foot on a deck. Many nautical terms have found their way into everyday usage, and the mere fact that I used the word "deck" without explaining it in the previous sentence is an indication of that.

What follows is a basic introduction to shipboard terminology. It will not make you an "old salt," but it will allow you to understand a lot more of the conversations around you and to participate in discussions with increased credibility. You are an important part of the Navy team, and proper terminology is one of the tools you deserve as part of your team equipment.

Structural Terms

Although the terms are different and there are certainly significant structural differences, a ship is in some respects like a building. There are floors (called "decks"), ceilings ("overheads"), corridors ("passageways"), and stairs ("ladders").

Walls are a little more complicated. Generally, you will be fine if you call what looks like a wall a "bulkhead," but technically a bulkhead is a wall that is structurally significant (supports decks and is watertight, for example). If a "wall" is only there to divide one space from another and is otherwise not structurally significant, it is more properly called a "partition."

What would be "rooms" ashore are generally called "compartments," but they are also called "spaces" (as in "engineering spaces") and sometimes even "rooms" (as in "fan room" or "wardroom"). Which word is used when does not lend itself to rules, except that either "compartment" or "space" can be used generically (as in, "A lot of those compartments need painting," or, "Most of those spaces up forward belong to first division"), and "room" is never used alone but is attached to another word as in the examples above.

Whereas ships have *storerooms*, and the items that are kept in them are generically called "stores," things are usually *stowed* rather than stored about ship, as in, "Those cables are stowed in the forward storeroom." And the word "stowage" is more nautical than "storage."

Doors are doors on ships just as in buildings, but there are technically two kinds. *Watertight doors* (WTD) are just what they sound like—doors that are specially designed to keep out water. They are strongly built and equipped with rubber gaskets around the edges to seal tightly when closed, and they have heavy-duty clasps called "dogs" to keep them tightly shut. *Non-watertight doors* (NWTD) are used inside the ship to separate compartments much as doors are used ashore.

One mistake you will see many people make—even Sailors—is to call a door a "hatch." The two are *not* synonymous: hatches refer to *horizontal* openings in ships that allow you to go from one deck to another; doors are *vertical* openings in bulkheads (or partitions) that allow you to pass from one compartment to another on the same deck. (To be really precise, the opening in a bulkhead or partition is a doorway and the object that closes it off is a door; similarly, the opening in a deck is a hatchway whereas the cover for it is the hatch [and it is sometimes called a "hatchcover"]. But you will often see "hatch" used as I have here, as in, "To get to the berthing compartment below, use that hatch over there.")

Windows (of which there are very few on most Navy ships) are usually called "ports," but you will also hear the term "bridge windows" used for those that are located on the bridge.

The uppermost deck that runs the entire length of the ship from bow to stern is the *main deck*. (An exception to this rule is the aircraft carrier, whose main deck is actually considered to be the hangar deck, not the flight deck, which would seem to fit the normal definition of main deck.) "Floors" below the main deck are called "decks," but those above the main deck are called "levels." (See "'Navigating' Aboard Ship" later in this chapter for a better understanding of these terms.)

The outside parts of a ship (which come in contact with wind and waves) are collectively known as the "skin" of the ship. The major part of the ship that is topped by the main deck is known as the "hull." Those decks that are exposed to the elements are called "weather decks."

To keep people from falling off the ship (overboard), ships have *lifelines* (made of strong cable) rigged all around the edges of the deck similar to railings on balconies ashore. If they are solid (instead of made of cable) they are more properly called "rails."

The structures above the main deck looking somewhat like houses or huts are collectively referred to as the "superstructure." Different kinds of ships have different types of superstructure; some may be one continuous mass, others may be split into separate groups; the superstructure projecting above the flight deck on an aircraft carrier is called the "island."

Often, the superstructure is topped off by one or more *masts*. At its simplest, a mast is a single pole extending vertically above the rest of the ship. Masts often are fitted with a horizontal crossbar, called a "yardarm," which is used to attach flag *halyards* (lines used to hoist the flags), or support navigational and signal lights and various antennas and electronic devices. If the ship has two masts, the forward one is called the "foremast," the after one the "mainmast." Modern ships do not normally have three masts, but in the days of sail, when masts also played a role in the propulsion of the ship by supporting her sails, some ships had a third mast, called the "mizzen," which was mounted after the mainmast. On single-masted ships, the mast, whether forward, aft, or amidships, is usually part of the superstructure and is simply called "the mast."

The *stack* of a ship serves the same purpose as the smokestack on a power plant ashore. It carries off smoke and exhaust gases from boilers on steam-powered ships and from gas turbines and diesel engines on ships with those forms of propulsion. Nuclear-powered ships do not need stacks because their reactors produce no smoke or gas. Some diesel-powered vessels release their exhaust from vents on their sides. On some ships, the masts and stacks have been combined to form large towers called "macks."

Ships are steered by a *helm* (also called a "ship's wheel" or simply "the wheel"). That wheel turns a *rudder*, the simplest design being a flat board or blade that extends into the water beneath the vessel's stern. In an open boat, a *tiller* can be attached directly to the rudder to turn it, but larger vessels have more complex steering systems (often including hydraulics, duplicate electrical cable systems, etc.) that culminate with the helm that is used by a helmsman to steer the vessel. Naval ships often have more than one rudder, but in the case of multiple rudders the rudders do not act independently but are controlled together.

Ships are driven through the water by propellers, and though this term is acceptable, you will more often hear them called "screws" (which avoids any confusion with aircraft propellers).

Ships or boats with two screws can be steered fairly well without a rudder by using the engines. If one screw turns faster than the other, the bow will swing toward the slower screw. If one screw goes ahead while the other goes astern, the bow of the ship will swing toward the backing screw; boats, and even very large ships, can turn within the diameter of their own lengths using this method, which is appropriately called "twisting."

Directions on a Ship

The front (forward) part of a ship is the *bow*; to go toward the bow is to "lay forward"; something that is located closer to the bow than something else is said to be "forward of" (as in, "Place that pallet forward of the others because it contains anchor parts"). The back (after) part of the ship is the *stern*; to go toward the stern is to "lay aft." Something located farther aft than another object is said to be "abaft" the other. You will also encounter the term "fore," which is usually used in conjunction with another word to indicate its relative position, such as the foremast; it is also occasionally used more generically as in, "Give the ship a clean sweep down fore and aft."

In a building you would go upstairs or downstairs; in a ship you "lay topside" or "lay below."* If you go extra high up in a ship, like climbing the mast, you would be "going aloft." If something is located high up on a ship (above the superstructure) it is said to be "aloft."

As you face forward on a ship, the right side is *starboard*, and the left side is *port*. Note that this does not mean that port and starboard are synonymous with left and right—if you are facing aft on a ship, starboard is now on your left and port is to your right.

Now it's time for some shipboard "relativity." Fortunately, you do not have to be an "Einstein" to understand it. We've already seen that objects are said to be forward or aft of each other depending upon their positions relative to the bow and stern—such as, "The mainmast is aft of the foremast," or, "The anchors are located forward of the superstructure." Now imagine a line running full-length down the middle of the ship from bow to stern; although you will never see one, it helps to know that it is theoretically there and is called, appropriately, the "centerline." An object that is closer to the centerline than another is said to be "inboard"; one that is farther from the centerline is said to be "outboard"—such as, "The lifelines are outboard of the mast," or, "The mast is inboard of the lifelines." This terminology is also sometimes extended beyond an individual ship, such as when ships are nested (moored alongside one another at a pier)—one closer to the pier than another is said to inboard, and vice versa.

Someone or something going from one side of a ship to the other is moving *athwartship*. This term also applies to something located in such a way as to run from one side to the other, as in, "An electrical cable has been rigged athwartship from the starboard side to the port." Something running in a

---

*Though the term "lay" is correct in this context, and you will hear it used, it is also acceptable to say "*go* topside" or "*go* below." You will find this preferred in the past tense: "He went below," or, "He went topside," *not*, "He laid below," nor "She laid topside."

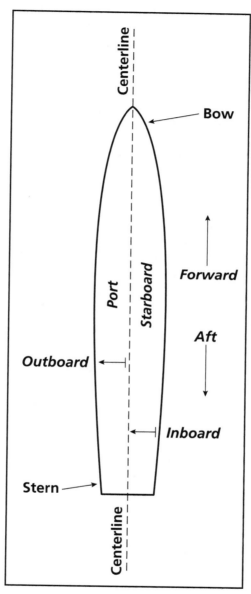

Figure 7.1 Directions on a Ship

direction perpendicular to this is said to run "fore and aft," as in, "Run a line fore and aft from the bow to the top of the mast."

Something located in the middle of the ship—neither forward nor aft—is *amidships*, as in, "Two of the inspectors headed for the bow, two others for the stern, and the rest remained amidships."

Underway and Other Conditions

Ships are considered *underway* when they are free-floating (i.e., not tethered to a pier or anchored to the bottom). You will see this term technically misused fairly often—even by Sailors—when it is used to describe a ship moving through the water, as in, "The ship was underway at fifteen knots." Though this statement is correct in that a ship cannot be moving through the water without being underway, confusion can result if one does not realize that a vessel can be underway without moving. Movement through the water is properly described as "making way." "The ship got underway at dawn and began making way at low speed once the signal to proceed was received," would be correct usage of these two terms.

In simplest terms, when ships are not underway, they are *moored* or *anchored*. Ships are moored to piers using *mooring lines*. To anchor, a ship uses her *ground tackle* (anchors and their associated equipment, such as chains and *windlasses* to haul them in).*

Important Locations Aboard Ship

The forward area of the main deck (about one-fourth to one-third the length of the ship, usually that area of the main deck forward of the bridge) is called the "forecastle" (pronounced *FOHK-sul*), and the after part is the "fantail" (usually the deck area aft of the superstructure).

In port, the *quarterdeck* is a formal area on a ship that serves as the point of entry or exit for people coming aboard or going ashore. It is manned by a watch team twenty-four hours a day and its location can change depending on how a ship is moored (it will be on the port side of the ship if she is moored port side to the pier, for example).

---

*There are some finer distinctions. A vessel that uses two anchors in a particular pattern is said to be moored, as is one that is attached to a mooring buoy. But these distinctions can generally be ignored with little consequence.

The *bridge* is the primary control station for the ship when she is underway, and the place where all orders and commands affecting the ship's movements and routine originate. It is roughly analogous to the driver's seat in a car or the cockpit of an aircraft, but it is large enough for a team of people to function as they steer, control the engines, watch radar scopes, talk on radios, etc.). For obvious reasons, the bridge is positioned such that it affords a good view of the outside world. The captain (commanding officer) will be on the bridge a lot of the time underway—especially during most special sea evolutions and when the ship is entering and leaving port—but obviously she or he cannot be there twenty-four hours a day. The OOD—a rotating watch position manned only by highly qualified personnel—is the captain's primary assistant in charge of safely running the ship and is always on the bridge (whether the captain is there or not) whenever the ship is underway.

The *combat information center* (CIC) is the nerve center of the ship, where information is collected, processed, displayed, evaluated, and disseminated to other parts of the ship (or to other ships) for use in decision making and in properly employing the ship. Sometimes it is known by other names (such as "combat direction center"). A wide range of electronic equipment is installed in CIC to process information received from a wide variety of sources, including radio, radar, sonar, electronic-warfare intercept receivers, IFF (identification friend or foe) transponders, visual communications, satellites, fathometers (depth gauges), and networked computers. CIC is the place where the ship's tactical operations are controlled. Such operations include the evaluation of targets, weapons firing, the control of friendly aircraft, surveillance operations, navigational assistance, submarine tracking, and many others.

The *engineering spaces* in a ship contain the propulsion machinery of the vessel as well as the various kinds of equipment that supply the ship with electricity (generators), fresh water (evaporators), and so on. Steam-powered ships will often have both boiler rooms and engine rooms, but gas turbine powered ships only have engine rooms. The primary control station for the engineering plant is usually called "main control."

Damage control central (DCC) serves as the central information and control site for matters affecting the safety of the ship. By monitoring conditions aboard ship and maintaining control of vital systems such as those used in firefighting and flooding control, and by maintaining careful records, damage-control charts, and liquid-loading diagrams, DCC sees that the ship is kept safe and stable during routine conditions and coordinates damage control operations when emergencies arise.

Magazines are special storerooms used for the stowage of missiles, rockets, and gun ammunition. For obvious reasons, these important but potentially dangerous areas aboard ship are kept locked and under close control. They also are protected by various alarm and firefighting systems and are usually located in spaces well below the waterline so that, in case of fire, they can be flooded.

The living spaces aboard ship are essential to accommodate the needs of the crew. Generally, enlisted crew members sleep in *berthing compartments*, whereas officers sleep in *staterooms*. The commanding officer's living area is called the "captain's cabin." Other living spaces include *heads* (lavatories), *wardrooms* (living and dining areas for officers), *messes* or *mess deck* (where enlisted personnel eat), and *galleys* (kitchens).* The medical clinic aboard ship is called "sick bay." A water fountain on a ship is called a "scuttlebutt." Larger ships may have other spaces for the health and comfort of the crew, such as tailor shops, libraries, barber shops, chapels, weight or aerobic rooms, and crew lounges. Virtually all ships have a *ship's store* that carries toiletries, uniform items, *gedunk* (snacks), and so on.

Shops and offices can be found on virtually every Navy ship. The number of each depends upon the size and the purpose of the ship. An aircraft carrier will have more than you can count; a patrol craft may have only one or two.

Vital Ship's Systems

A number of systems are essential to every Navy ship and, in many ways, are analogous to the nervous, circulatory, respiratory, and excretory systems of the human body.

The electrical system is essential for running complex weapon and communications systems, computing the solutions to a vast spectrum of tactical problems, powering ammunition hoists and aircraft elevators, detecting incoming enemy missiles and aircraft, running in-house television systems for entertainment, and so on. These, and hundreds of other functions, make electricity as vital to a modern vessel of war as ropes were to a sailing vessel.

---

*On larger ships, you may encounter multiple messes. Besides a chief's mess (for E-7 through E-9), there may be a first-class mess (for E-6s) and possibly a warrant officer's mess.

Ships generate their own electricity and all have backup systems to provide power when the primary system fails. Vital electrical circuits are also frequently duplicated so that power can continue to flow after battle damage occurs.

The ventilation system supplies fresh air where it is needed and carries off unwanted exhaust. Supply ventilation brings fresh (external) air into the ship and, in the event of cold weather, heats the air by means of preheaters installed in the ducting. Exhaust ventilation carries away the air that has served its purpose and needs to be replaced. In those spaces containing equipment that generates heat or humidity or both (such as main engineering spaces, galley, or head facilities), the exhaust system is particularly vital. "Recirc ventilation" is provided to spaces containing electronic equipment (which requires a cool environment for proper operation), as well as to berthing, messing, and office spaces. As its name implies, this system recirculates internal air to prevent stagnation and, when necessary, draws the air through a cooling system to maintain the proper temperature.

In the event of fire, flooding, or some other danger requiring the isolation of a space or spaces, ventilation systems can be secured by de-energizing the fan motor and can be segregated by closing valve-like devices in the ducting (often found where the ducting penetrates decks, overheads, and bulkheads).

The potable water system provides water for drinking, personal hygiene, and cooking. Potable water is made in the ship's distilling units (evaporators) from saltwater taken from the sea, and it is stowed in tanks specifically designated for potable water only. Piping systems carry the water from the tanks to the heads, galleys, and drinking fountains (scuttlebutts) where it can be used.

The saltwater system provides other water needs. Drawn directly from the sea through underwater intakes and pumped throughout the ship using a different piping system from the one used for potable water, this water is available for firefighting when needed and is used on a routine basis as flushing water for the heads.* Saltwater is also is used as cooling water for certain items of machinery and electronic equipment and can be piped into tanks for ballast (to stabilize the ship). Special sprinkler heads mounted all over the outside of

---

*If you would like to see proof of the latter, try turning the lights off in the head, and then flush a toilet. You may well see phosphorescent lights swirling in the water—these are tiny bioluminescent animals that have been brought into the ship through the sea intakes that supply the saltwater system.

the ship can be opened to allow a wash down of the ship to rid her of contaminants in the event of a chemical, biological, or radiation (CBR) attack.

The drainage system includes the piping, valves, and pumps that discharge water from the ship. It is used routinely for pumping a ship's bilges (the lowest parts of the ship's hull where water collects) and in emergencies to remove seawater that has entered the hull because of damage, collision, or heavy weather. It is also coupled with the saltwater system so that water can be shifted around to maintain proper trim (stabilize the ship).

Because weather-deck drains collect natural rain and seawater, the drains connected to these areas are piped directly overboard. But internal drains (from sinks, showers, galleys, toilets, and urinals) are carefully controlled for environmental reasons. Drainage from these sources is collected in specially designed tanks for appropriate disposition.

The fuel system includes fuel-stowage tanks, pumps, filling lines, transfer lines, and feed lines to the ship's boilers, engines, and generators. Like the other liquid systems aboard ship (potable water, saltwater, and drainage) the fuel system is also constantly monitored and moved about to help maintain proper trim (stability).

The compressed air system is used to charge torpedoes, operate pneumatic tools, run messages through pneumatic dispatch tubes, power automatic boiler controls, eject gases from guns after they have fired, and so on. Compressors create the compressed air and special piping carries it where it is needed in the ship.

GOING ABOARD

Just as it is a good idea for people working for the State Department to learn something about a foreign country before traveling to it, so should you learn something about the somewhat foreign world of ships before going aboard. In the previous chapter and the earlier sections of this one, we have learned about the basic "geography" and language of ships. Now we will learn some of the customs, cautions, and practices that you will need to understand to have a safe, enjoyable, and worthwhile experience while embarked in a Navy ship.

Arrival

First, some clarification. You have probably encountered the terms "aboard" and "on board" and wondered what the difference is. There isn't much. Both are acceptable. Although "aboard" is usually preferred, you can also use "on board" and be safe. Do not use "onboard" (as one word) when writing, however; that *is* incorrect in naval usage. You will see it hyphenated as an adjective, however, as in, "We will be using the on-board computers for our calculations." One convention (but not a rule) is that *aboard* deals with people and *on board* deals with equipment, as in, "Be sure to take the laptops on board next week when we go aboard the ship."

You will also encounter the word "embark" (as in the above paragraph), but it is generally used to indicate people (usually not equipment) who are not part of the regular crew, as in, "The reporter will embark next week to interview those Sailors still aboard who took part in the operation," or, "The Marines will embark at dawn for transport to the assault area." Units that contain people can also embark, as in, "The battalion has been embarked for most of the deployment."

One more distinction. Crew members serve *in* a ship, not "on" (as in, "She served in USS *Independence* for three years"—*not* "She served on USS *Independence* for three years").

The opposite of aboard is ashore. The opposite of embark is debark or disembark. Webster's prefers the latter; the Navy tends to prefer the former.

When you go aboard a ship, you will probably cross from a pier to the ship and you will use the brow (not the "gangplank") to cross over. If you are going aboard a ship that is anchored out, you will come aboard from a boat using an accommodation ladder rather than a brow. In either case, you will first encounter the equivalent to an entrance hall or foyer in a building; this is called the quarterdeck (already mentioned earlier). This may be an open deck area or it may be covered by an awning or actually located just inside the skin of the ship. The quarterdeck is considered a formal, sometimes even ceremonial area. It is not a place for loud talking or laughing. You might note that Sailors coming aboard or going ashore via the quarterdeck go through a ritual of saluting both the national ensign (located at the ship's stern) and the quarterdeck watch officer and reporting his or her intentions to the latter very formally. You also should report to the quarterdeck watch officer (often called the "officer of the deck" or "OOD"). You will be required to show proper identification and any other relevant paperwork (orders, for example).

Keep in mind that boarding any Navy ship is similar to entering a Navy installation ashore and you are giving full consent to a search of your bags and baggage just by being there. Although baggage inspection is not always done, be prepared to submit your baggage if requested to do so.

What to Bring

Keep in mind that, though help may be available, you may have to carry whatever you bring up and down rather steep staircases (ladders) and through narrow passageways. Distances on a large ship, such as an aircraft carrier, can be quite long and, on an attack submarine, passages can be particularly cramped. Also bear in mind that stowage space for your belongings will not be spacious. So don't bring more than you will need.

Though the Navy prides itself on keeping its ships as clean as possible, it is also a place where lubricating oils and other substances not friendly to nice clothing abound, so durable, washable clothing is a better choice over your favorite special-care clothing. If you will be embarked for a longer period and must subject your clothing to the ship's laundry, remember that your clothes will be in rather rough company and delicate treatment is not the order of the day—though only legendary, the "button crusher" that is supposedly standard equipment in every ship's laundry did not spring entirely from fertile imaginations!

Odds are you will have to travel some distance thorough a passageway or two to reach the nearest head, so be sure to bring a robe of some kind for the trek to the shower. Also bring some sort of shower shoes (flip-flops or the like) to prevent stubbed toes and a case of athlete's foot.

You never know when you might be rousted from your bunk in the middle of the night (in case of a drill or emergency), so it is better to sleep in something you wouldn't mind (too much) being seen in as you stumble half-asleep to an assigned mustering station.

Many ships have some form of exercise equipment, so athletic gear is not a bad idea if you are going to be aboard long enough to make it worthwhile. There is nothing quite like a jog on an aircraft carrier's flight deck if you are fortunate enough to be aboard during one of those times when flight operations are not taking place.

Cameras are usually acceptable and are recommended if going aboard is a new experience for you, but follow instructions and use discretion in shooting photographs. If in doubt, ask or don't shoot.

If you bring a compact disk player or the like, keep it small, bring extra batteries (they may or may not be available in the ship's store), and don't count on a lot of available electrical outlets. Be careful about "tuning out" the ship entirely through the use of headphones and loud music—important announcements can be made at any time.

A small flashlight is a good idea. But make sure it has a red lens, or be very careful of shining it near nighttime watchstanders (you can temporarily remove their night vision with white light).

"Navigating" Aboard Ship

Finding your way around a patrol craft will not prove too challenging, but people have been known to get lost on aircraft carriers for years (slight exaggeration, but the point is made). Fortunately, crew members are usually more than happy to show off their knowledge and help you find what you are looking for. But there is also a system by which you can find your own way around—and it is logical, the same for all Navy ships, and not hard to learn.

Just as a town or city has a system using street signs and addresses to help you find your way around, so does a Navy ship. Each compartment on a ship has an identifier (actually posted on the bulkhead or partition in each compartment) that is roughly equivalent to a street address in a city. Once you understand this system, you will know where you are at any given time on a ship, and you will be able to find any space on a ship even if you haven't been there before.

Each and every space in a ship has a unique identifier made up of numbers and letters, such as:

$$4\text{–}95\text{–}3\text{–}M$$

For each compartment, the numbers and letters will change, but the format will not. That format tells you the following:

- First number = deck or level number
- Second number = frame number
- Third number = compartment's position relative to the centerline
- Letter(s) = code telling you the purpose of the compartment

To understand what those numbers and letters are telling you, a review of some basic ship construction is useful.

You will recall that we earlier established that ships have a main deck. That main deck is one of the references used in the compartment numbering system, and all "floors" above the main deck are called "levels" and all below are called "decks." These are indicated by the *first number* of the compartment

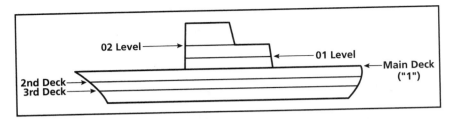

Figure 7.2   Deck Numbering System

identifier. Some important clues to deciphering this number system are in knowing that the main deck is always numbered "1," that all decks below the main deck are numbers higher than "1" (2, 3, etc.), and that all levels above the main deck are preceded with a zero (01, 02, etc.). These numbers increase as you go away from the main deck. The first deck below the main deck is numbered "2," the next one down is "3" and so forth. One level above the main deck is the "01" level (conventionally pronounced "oh-one level," not "zero-one level." The next one up is the "02 level," and so on. If a compartment extends through more than one deck (such as an engineering space that must be large enough to hold a boiler or huge turbines), the deck number of that compartment refers to the bottommost deck.

It also helps to know that shipbuilders start with a virtual backbone, called a "keel," that is essentially a *large* I-beam running from bow to stern. Other beams, called "frames," are attached at regular intervals to that backbone—roughly perpendicular (athwartships) to it—like ribs to the human spine. Plating is attached to these ribs to form the ship's hull. These frames are numbered, starting at the bow and increasing toward the stern—the farther aft you are, the higher the number of the frame nearest you. This is another clue to navigating your way around the ship, because the *second number* of the compartment identifier tells you the frame number. It's even more useful to you if you know how many frames the ship has in total; that way, if the ship has three hundred frames, for example, and you are at frame one hundred, you know you are about one-third of the way aft on the ship (or two-thirds of the way forward). Frame 150 in this case would be exactly amidships.

The *third number* is referenced to that imaginary line we described above—the centerline—which runs from bow to stern, bisecting the ship into two long halves. This third number in the compartment identifier tells you where you are in relation to that centerline. The bigger the number, the farther away from the centerline (outboard). Compartments with even numbers are on the port (left, looking forward) side of the centerline; odd numbers are on the starboard (right, looking forward) of the centerline.* This means that a compartment with the number "3" in this position on the identifier would be the second compartment outboard from the centerline ("1" being the first) on the starboard side. If you understand this system, you will know that the second compartment on the port side would have the number "4," the third compartment on the port side would be "6," the third on the starboard side would be "5," and so on. If a space straddles the centerline, it has a zero as its third number.

The *last part* of the compartment identifier does not tell you much, if anything, about the location (although, if you are looking for an office and the compartment you are standing in front of has a code letter "V," you are probably lost). This letter, or pair of letters, is a code that tells you what the space is used for. Though you don't really need to know these codes to be able to get around on ship, they can be useful for other purposes and are provided here:

A	supply and stowage spaces
AA	cargo holds
C	control centers (such as CIC)
E	engineering (machinery)
F	fuel stowage (for use by the ship itself; that is, not as cargo)
FF	fuel stowage compartments (when cargo)
J	jet (aviation) fuel for use by embarked aircraft
JJ	jet (aviation) fuel as cargo
K	chemicals and dangerous materials (other than oil and gasoline)
L	living spaces (berthing and messing spaces, heads, passageways, etc.)
M	ammunition spaces (magazines, ready service lockers, etc.)
Q	miscellaneous spaces not covered by other letters
T	vertical access trunks or escape trunks

---

*Numbering systems in the Navy—whether for compartments as in this case, or for navigational buoys you encounter as you come in from sea—usually assign even numbers to the port (or left) side and odd to starboard (or right). A way that may help you to remember this is that the words "left," "port," and "even" all have four letters—works for me, anyway.

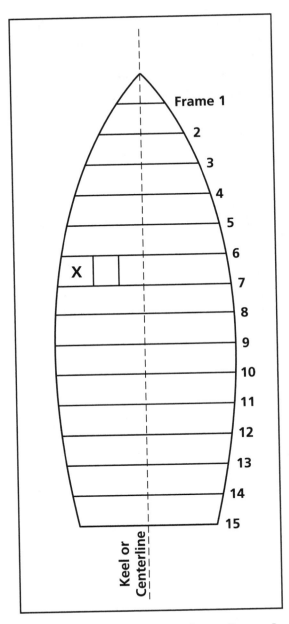

Figure 7.3 Compartment Numbering System. In the example above "X" is in compartment 4-6-6-M (fourth deck, sixth frame, third compartment to port, which is a magazine).

V   void (spaces that are normally empty)

W   water stowage spaces

So, in the example above (4–95-3-M), the compartment identified is on the fourth deck, is at the ninety-fifth frame, is the second compartment outboard from the ship's centerline, and is a magazine or some other ammunition-related space.

Let's suppose you are in this space and on the ship's 1MC (pronounced "one-em-see"—the general announcing system) you are asked to report to the ship's CIC, which has a compartment identifier of 03–47-0-C. To get there from this ammunition space, you would know that you would have to go up six ladders (three to get to the main deck [number 1] and three more to go up to the 03 level). You would also know that you had to go forward rather than aft, because the frames always go up in number as you go aft and down in number as you go forward. If you did all that and arrived in a space numbered 03–47-5-C, you would know that you were close but that you had to head inboard (away from the starboard side toward the centerline) to find CIC (which you know from the "0" is on the ship's centerline). Using this system, you can find any space on any Navy ship.

As mentioned earlier, compartment identifiers are actually posted in each compartment on a ship (sometimes more than once in larger compartments), so you can always have a sense of where you are in a ship. These identifiers are painted on and are easily found. They all follow the same format as follows:

4–95-3-M

FR 95–99

GM

The top line is the one we have been discussing, the one that tells you where you are on the ship. The second line tells you how many frames the compartment spans, and the bottom line tells you what division on the ship is responsible for maintaining that space (in this case, GM Division, the people who take care of the guided missiles on the ship).

Emergency "Navigation"

Though in many ways, you will be safer aboard ship than you are in your own home, bad things do sometimes happen aboard ship. Fire is a more serious hazard aboard ship than even flooding, and you should take precautions to be sure you are ready in the event one breaks out while you are aboard.

Because ports (windows) are rare, ships can be *very* dark places if power goes out; smoke can also quickly accumulate, making it very difficult to see. As soon as you find out where you will be sleeping, you should plan an emergency escape route (or more than one if possible). Do the same with any other spaces where you will be spending a lot of time. Make several trips to the nearest passageways that will take you to safety, until you become very familiar with the route and can literally travel it with your eyes closed. Practice doing it with your eyes closed so that you will be ready in the event of an emergency. I lived for two years near the engineering spaces in USS *Saratoga* where three men had several years before died during a fire. To preclude that happening to me, I closed my eyes *each and every time* I left my stateroom so that I was more used to doing it that way than with full vision.

You will be told what to do in emergencies. You may be assigned a place to muster (report your presence) in the event of certain emergencies. For example, if someone is thought to have fallen overboard, a quick muster of all hands on board is essential to determine if someone is indeed missing. Be sure that you know where you are supposed to muster and that you know how to get there as expeditiously (and safely) as possible. Consider alternate routes where possible in the event a fire or something else is in your path.

Remember the term "FUSPAD"! This is an acronym that means:

Forward—*Up*—Starboard
Port—*Aft*—Down

This tells you the flow of traffic on a ship during an emergency. If the ship goes to General Quarters (GQ) (you will know when that happens when you hear a loud "bonging" alarm, the thunder of many footsteps, and the words, "General Quarters, General Quarters. All hands man your battle stations," and so on, over the ship's 1MC), everyone will be moving at once to get to their emergency stations. If there were not a "traffic pattern," Sailors would collide in narrow passageways and block ladders. (I have known some gunner's mates you would *not* want to collide with.) So everyone aboard ship knows that in emergencies, they must use passageways on the starboard side of the ship if they need to go forward, and they must use ladders on the starboard side if they need to go up to a higher deck or level. Conversely, those who need to move aft or to lower decks must use passageways and ladders on the port side of the ship.

Visiting Special Areas

One of the most popular places to visit a ship is her bridge. If you are fortunate enough to be invited to the ship's bridge, be sure to do the following:

- Be aware that the watchstanders have priority at all times. Do not get in their way if they are trying to move about, and do not block their view if they are trying to see out or view a computer display.
- Do not speak too loudly. Watchstanders must be able to hear one another, radios, etc.
- Do not use binoculars or any other equipment unless invited to do so.
- If you are invited to the bridge at night, be sure to use only a red-lensed flashlight and be very careful where you shine it (mostly on the deck).

If you are visiting an aircraft carrier, it should be pretty obvious that you should never venture onto the flight deck or the surrounding catwalks during flight operations unless specifically invited and escorted. People who work on the flight deck receive hazardous duty pay for a reason. However, you will not want to miss the opportunity to observe flight operations while you are aboard. There is a place on every carrier's island (on the 09 or 010 level) called "Vulture's Row," where you can actually go outside and have a "balcony" view of the "ballet" of Sailors and aircraft below. Be sure to wear ear protection and remove your hat and anything else (like objects in your shirt pockets) that might be blown on to the flight deck. Cameras are usually fine but *never* use a flash.

Some Final Thoughts on "Navigating"

Before proceeding up or down a ladder, be sure to look. Because ladders are narrow and steep, you might catch a knee in the face if you start up when someone else is coming down. Sailors often move quickly on ladders, sometimes swinging from the handhold above and deliberately missing the last few steps altogether, so *be vigilant*.

Beware of the infamous "knee-knockers." Doorways that pass through bulkheads are designed to prevent the free flow of water for rather obvious reasons, so the opening does not always reach all the way to the deck. The bottom may be elevated such that you have to step over in order to go through. Obviously, if you are not paying attention, you could well stub a toe or bang a knee as you go through this restricted opening. Often the top of the opening is also lower than a regular door ashore, so you will do well to be alert and *duck* when going through.

Hatches also have a raised edge (called a "coaming") around the opening so that water cannot readily flow down through to a lower deck. Be careful to step over this when going up or down a ladder.

If you encounter tape down the middle of a passageway, it is probably because someone is swabbing (mopping) or waxing the deck. Work is done on one half at a time, so that half the passageway is still available for passage. Be sure to stay on the correct (dry) side when proceeding.

Telling Time

Because the Navy frequently uses twenty-four-hour time ashore as well as aboard ship, that system is covered in another chapter of the book (see chapter 10). You will do yourself no favors by going aboard without understanding how to tell time this way.

There is another form of timekeeping you may encounter aboard ship, particularly larger ones, that bears some explanation. You may hear bells ringing in a routine but strange pattern. This is more a matter of tradition than of practicality, but it is one of those things that make shipboard life interesting, even a bit exotic.

For many centuries, Sailors did not have the luxury of a personal timepiece. If watches were to be relieved on time, some means of telling the time had to be devised. A system that used a half-hour sandglass and the ship's bell was created and used for hundreds of years.

At the beginning of a watch, the sandglass was turned over to start it running. As soon as it ran out, the watchstanders knew the first half hour had passed, so they rang the ship's bell once and immediately turned the sandglass over to start the second half hour. Everyone on board the ship could hear the bell, so they could keep track of the time. When the sand ran out the second time, the watchstanders rang the ship's bell twice. They continued this until eight bells had been rung (representing the passage of four hours, or one complete watch). The watch was then relieved, and the new watch team started the whole cycle over by ringing one bell when the first half hour had passed, and so on. This bell-ringing tradition has been continued on board many Navy ships even though clocks and watches are now very common.

Even though this system is archaic, you can actually tell time by it (within a half hour). Remember that watches are four hours long and that noon and midnight are among the starting points for watches; the rest is a matter of extrapolation. If a watch begins at noon, then you will hear one bell at 1230, two bells at 1300 (1:00 pm), three at 1330 (1:30 pm), and so on. The cycle will end at the four-hour point (1600 or 4:00 pm) with eight bells and begin again with one bell a half hour later (1630 or 4:30 pm).

Today, because bells are rung more out of tradition than for real function, they are not normally rung between taps and reveille (normal sleeping hours for Sailors not on watch), nor are they rung during divine services or in fog, when the ship's bell is used as a fog signal. Another tradition, still observed in many Navy ships, is the custom of the youngest member of the crew striking eight bells at midnight on New Year's Eve to ring in the New Year.

Dos and Don'ts

Do read the Plan of the Day (POD). The ship's Plan of the Day comes out the night before the day it covers and lists information pertaining to the coming day's routine, such as what special evolutions or drills are scheduled. It is a good source of information and reading it will reduce the number of questions you will have to ask. It may contain information you might not want to miss, such as a movie schedule.

Don't take long showers. Because the evaporators can make only so much water at a time, care must be exercised not to waste fresh water while a ship is underway or at anchor. You should never take what is popularly called a "Hollywood shower" while at sea. (This is the kind of shower you probably take at home, where you let the water run for as long as you are in the shower.) While at sea, it is important to get in the habit of wetting down (quickly), turning the water off and leaving it off while you soap and scrub, and then turning the water on again just long enough to rinse off.

When moored to a pier where the appropriate connections are available, the ship's potable water system can be hooked up to receive fresh water. At these times, abundant fresh water is available so that the strict water conservation practices you use at sea are not necessary. The ship's potable water tanks will all be topped off (filled to capacity) before the ship gets underway.

Do secure doors after passing through. If you have to spin a wheel, work a large lever, or turn a series of small levers in order to go through a door, then you should turn that wheel or work those levers in the opposite direction after you have passed through. Those wheels and levers are different methods you might encounter for securing (tightly closing) a watertight door (to *keep* it watertight).

Don't open any doors without permission if the ship is at GQ. The ship is "buttoned up" during GQ to prevent fire, smoke, fumes, or water from passing from space to space. To breach that seal without permission is a sin of high order.

Don't move about during a Security Alert. If you hear a Security Alert called away while you are aboard, you should stand fast. The only moving about you should do is to get out of the middle of a passageway or off of a ladder. Armed Sailors and/or Marines will be moving about the ship rapidly and will be in a no-nonsense mode. Do what they tell you—they will not stand on ceremony!

Don't go beyond a RADHAZ sign. You may encounter special signs with the term "RADHAZ" on it, particularly up high in the ship. This indicates a radiation hazard (usually from a radio or radar antenna) and you will do well to heed it. To ignore it can bring you serious injury and even death.

Do stay clear of emergency sites. In the event of an emergency, such as a fire, the word will be passed over the ship's 1MC, "Fire, Fire, Fire," along with the location (another reason to know how to decipher the compartment identification system). You should stay well clear of this area and out of the way of crew members who are proceeding to the area.

Do remove headgear (hat, cap, helmet) when entering or traveling through a shipboard eating area (mess deck, chief's mess, wardroom, etc.). A Sailor on watch (such as a messenger) will keep his or her headgear on when entering these spaces, but all others do not.

Do be careful of wet paint. Although modern ships are less vulnerable to rust and other forms of corrosion, the sea is a hostile environment to metal. Consequently fresh paint is a fairly common occurrence, so be careful what you touch or lean up against.

Do review the chapter of this book on Navy customs and traditions. There are some things in that chapter (such as passing honors between naval vessels) that you should be familiar with.

Do enjoy the experience. Few people ever venture onto the high seas, and many fewer still in a warship. Make the most of the experience (ask questions, observe, etc.). You will likely speak of it for many years to come.

QUICKREFS

- The following terms are roughly equivalent (and you may hear many of them used ashore in Navy buildings as well as aboard ship):

  floor = *deck*
  ceiling = *overhead*
  wall = *bulkhead* (sometimes *partition*)
  room = *compartment* (or *space*)
  corridor or hall = *passageway*
  stairs or stairwells = *ladder*
  drinking fountain = *scuttlebutt*
  lavatory (or bathroom) = *head*
  window = *port* (although *bridge window* is acceptable)
  outer walls = *skin*
  safety rails = *lifelines*

- Doors are doors (but they may be *watertight* or *non-watertight* and the former will have gaskets and *dogs* to seal them tightly).
- A *hatch* is an opening in a deck that allows vertical access to other spaces.
- *Masts* are essentially poles used to hold flags, antennas, and so on aloft.
- A crossbar on a mast is called a *yardarm*.
- Ships are steered by the *helm*, which turns a *rudder* in the water.
- Propellers drive most ships through the water and are often called *screws*.
- The *main deck* is the uppermost deck that runs continuously from bow to stern (except on aircraft carriers where the hangar deck and not the flight deck is the main deck).
- "Floors" below the main deck are called *decks*; those above are called *levels*.
- Directions or relative positions on a ship are as follows:

  entering a ship = *go aboard* or *on board*
  front end of a ship = *bow*

back end of a ship = *stern*
to move toward the bow = *lay forward*
located closer to the bow = *forward*
to move toward the stern = *lay aft*
located closer to the stern = *abaft*
to go upstairs = *lay (or go) topside*
to go downstairs = *lay (or go) below*
located upstairs = *topside*
located downstairs = *below*
to climb a mast = *go aloft*
located above the superstructure = *aloft*
right side of ship when facing forward = *starboard*
left side of ship when facing forward = *port*
imaginary line down the middle of the ship from bow to stern = *center-line*
direction from centerline toward the outer side of the ship = *outboard*
direction from side of ship toward the centerline = *inboard*
direction going from one side of a ship to the other = *athwartship*
something running lengthwise on a ship = *fore and aft*
the middle part of a ship lengthwise = *amidships*

- If a vessel is *underway*, it means she is free-floating (and may or may not be moving).
- A vessel that is *making way* is moving through the water.
- Generally speaking, vessels not underway are *moored* (tied to a pier) or *anchored* (to the bottom).
- Important locations aboard ship:

forward area of main deck = *forecastle*
after area of main deck = *fantail*
formal entryway to a ship = *quarterdeck*
primary control position for a ship when underway = *bridge*
nerve center for managing information = *combat information center (CIC)*
location of propulsion and auxiliary machinery = *engineering spaces*
safety control center = *damage control central (DCC)*
ammunition stowage spaces = *magazines*

enlisted living and sleeping areas = *berthing compartments*
officer sleeping quarters = *staterooms*
lavatories = *heads*
enlisted dining area = *mess deck* or *mess*
officers' dining area = *wardroom*
kitchen = *galley*
medical clinic = *sick bay*

- Vital systems found aboard Navy ships include electrical, ventilation, potable water, salt water, drainage, fuel, and compressed air.
- *Aboard* is generally preferred to *on board*. *Onboard* is not a Navy word. *On-board* is an acceptable adjective (to describe equipment, etc.).
- People (or groups of people) who are not regular crew members *embark* (go aboard) and *debark* (go ashore).
- Crew members serve *in* a ship (not "on").
- If you are fortunate enough to be invited aboard a Navy ship, pack as little as possible; stowage is tight. Don't forget to bring durable clothing, shower shoes, a robe, and athletic gear (if you are going to be aboard for a while). You may bring a camera if you like but use it with discretion (if in doubt, ask or don't shoot). Another useful item is a flashlight (one with a red lens is best).
- Every compartment has a unique "address" that makes it possible to find it. It will look like this: "4–95-3-M."

First number = deck or level number
Second number = frame number
Third number = compartment's position relative to the centerline
Letter(s) = code telling you the purpose of the compartment

—The *first number* (4 in this example) tells you the deck or level (fourth deck in this case). Decks are either the main deck (numbered "1") or below (numbers get higher as you go down). Levels are above the main deck (the numbers get larger as you go up) and they have a zero preceding the number (such as "04" or "012").
—The *second number* (95 in this example) tells you the frame number—how far forward or aft you are. The first frame up forward is "1" and they increase as you head aft.

—The *third number* (3 in this example) tells you where you are in relation to the centerline of the ship. Compartments are numbered 1, 3, 5, and so on (odd) as you go away from the centerline on the starboard side. Compartments are numbered 2, 4, 6, and so on (even) as you go away from the centerline on the port side.

—The *last letter(s)* tells you the purpose or usage of the compartment. Different letters have different meanings (M for magazine or some other munitions-related purpose).

- For emergencies (such as fire or man overboard):
  —You should learn escape routes (with your eyes closed) in case of emergency.
  —Know where you are to muster (report) in case of an emergency.
  —Remember *FUSPAD* for how to travel on a ship in an emergency (*For*ward and *U*p on the *S*tarboard side, *P*ort side for *A*ft and *D*own).

- Be unobtrusive and quiet if visiting the bridge. Do not touch equipment unless invited.
- Never go onto a carrier's flight deck without invitation and escort.
- "Vulture's Row" is a great place to observe flight operations on a carrier.

  —Wear ear protection.
  —Do not wear a hat or cap.
  —No loose items in your pockets.
  —Do not use flash on your camera.

- Be careful of
  —Sailors coming down ladders,
  —"knee knockers,"
  —low doorways,
  —hatch coamings, and
  —wet decks where you see tape dividing a passageway.

- Learn military (twenty-four-hour) time before going aboard.
- Ship's bells actually tell you the time (within a half hour).
- Read the Plan of the Day for useful information about the ship's routine.

- When showering at sea turn the water off while soaping. Use as little water as possible.
- Close doors behind you.
- Do not open any doors during GQ (General Quarters) without permission.
- During Security Alerts, stand fast (stay where you are)—unless you are in the middle of a passageway or on a ladder.
- Do not go past a RADHAZ sign without permission.
- Remove hats or caps when entering a mess (eating area).
- Watch out for wet paint (unless you like gray trim on your clothes).

CHAPTER EIGHT

Aircraft

Naval aircraft are an essential component of sea power. The U.S. Navy has thousands of aircraft in its inventory, performing a wide variety of missions, many from the decks of ships and others from naval air stations all over the world. The many kinds of fixed-wing and rotary-wing (helicopter) aircraft flown by the Navy include fighters, attack, combined fighter-attack, anti-submarine, patrol, early warning, general utility, in-flight refueling, transport, and trainers.

Like ships, aircraft have their own specialized vocabulary. Included here is enough basic terminology to allow you to converse and understand the language of this specialized field.

Naval aircraft are organized into squadrons and these are further grouped into air wings.

BASIC AIRCRAFT NOMENCLATURE

Because aircraft are such an important component of the Navy, familiarity with certain basic terms concerning the structure of airplanes and helicopters can be useful.

The *fuselage* is the main body of the aircraft. The wings are strong structural members attached to the fuselage. Their airfoil shape provides the lift that supports the plane in flight. Wings are fitted with controllable *flaps* for increased lift and may be fitted to carry guns, rockets, missiles, and other weapons, as well as fuel tanks, engines, or landing gear. Besides fixed-wing aircraft, there are "rotary-wing aircraft," more commonly called "helicopters"; instead of having wings in the traditional sense, they have *rotors* (which are actually wings that rotate).

The tail assembly of a fixed-wing aircraft usually consists of vertical and horizontal stabilizers, rudder(s), and elevators. These components are key elements in the flight controls of the aircraft. Helicopters often have a tail rotor that keeps the helicopter from spinning like a top.

The "landing gear" usually means the wheels, but in certain specialized aircraft these may be replaced by skids, skis, or floats.

Fixed-wing aircraft that land on aircraft carriers have *tail hooks*, which are designed to catch a cable stretched across the flight deck, allowing the aircraft to land in a very short distance.

The *powerplant* develops the thrust or force that propels the aircraft forward, providing mobility and—in combination with the wings—the lift necessary to keep the aircraft aloft. In the case of helicopters, the powerplant provides the power to keep the rotors spinning, which keeps the aircraft aloft and allows it to hover as well as move through the air. The powerplant may consist of reciprocating (piston) engines that drive propellers, jet engines that develop thrust (turbojet and turbofan), or turbine engines and propellers or rotors in combination (turboprop or turbo shaft).

Another useful aircraft term is "Mach," which is commonly used to measure the speed capability of an aircraft or missile. Formally defined as the ratio of speed of an object to the speed of sound in the surrounding atmosphere, it is used as follows. An aircraft traveling at Mach 1 would be moving at the speed of sound. One going Mach 2 would be going twice the speed of sound, and Mach 1.5 would be one-and-a-half times the speed of sound. Depending upon the altitude, temperature, and some other variables, the speed of sound varies, but a rough figure to use for approximation is 650 miles per hour. So an aircraft flying at Mach 2 would be moving at a speed of approximately thirteen hundred miles per hour. An aircraft that is able to fly faster than the speed of sound (Mach 1) is said to be "supersonic" and one that cannot is called "subsonic." A "hypersonic" aircraft flies at Mach 5 or greater.

Stealth technology has changed the appearance and actual structure of aircraft considerably. Stealth is accomplished by incorporating different materials and designing an aircraft's outer surface at various angles that will deflect a radar signal rather than return it to the transceiver from which it came, thereby making the aircraft virtually invisible to enemy radar.

TYPES OF NAVAL AIRCRAFT

There are many different types of aircraft in the U.S. Navy's inventory (Photos 8.1 through 8.6). Some of these were designed specifically for naval use, but many are used by the other armed forces as well. Some are fixed-wing, whereas others are rotary-wing (helicopters).

Fighters are used to destroy other aircraft and incoming missiles. They are the aircraft you would normally see involved in a "dogfight." Fighters are very fast and highly maneuverable. They patrol above friendly forces in what are called "combat air patrols" (CAP) and intercept and engage incoming enemy aircraft or missiles. They also penetrate enemy air space to engage the enemy's aircraft and escort other kinds of aircraft when they are carrying out their missions in hostile areas. All fighters are fixed-wing.

Attack aircraft are designed to destroy enemy targets, at sea and ashore, such as ships, vehicles, transportation systems, airfields, enemy troops, and so on. To accomplish these missions, attack aircraft are armed with various configurations of rockets, guided missiles, gun systems, torpedoes, mines, and bombs. Attack aircraft can be either fixed-wing or helicopters.*

Patrol aircraft are tasked primarily with finding enemy forces. They are designed more for long range and time on station than for speed. Although they may be armed, sensors (such as radar, infrared, acoustic, and magnetic-detection devices) are their most important components.

Antisubmarine aircraft search out submarines visually, by radar and magnetic detection, or by signals sent from floating sonobuoys, and then destroy them with rockets, depth charges, or homing torpedoes. Both helicopters and fixed-wing aircraft are used for antisubmarine warfare.

Mine-warfare aircraft lay mines in enemy waters or sweep enemy mines from friendly waters or objective areas. Both rotary- and fixed-wing aircraft can be used in mine warfare.

Command and control aircraft coordinate various operations within the battle space using sophisticated sensors, communications, and computer equipment.

Electronic warfare aircraft are designed and built specifically for tactical electronic warfare operations, such as jamming enemy radars for a significant tactical advantage.

Transport aircraft are used to carry cargo and personnel. As with any of the types of aircraft in the Navy's inventory, some are land-based and others can be operated from aircraft carriers. Both helicopters and fixed-wing aircraft are used for transport missions.

---

*Some aircraft are designed to carry out both fighter and attack missions.

Photo 8.1   F/A-18 Super Hornet. This aircraft is a combination fighter and attack aircraft.

Photo 8.2   The S-3B Viking is an antisubmarine aircraft.

Photo 8.3   The E-2C Hawkeye is a type of command and control aircraft.

Photo 8.4   The EA-6B Prowler is an electronic warfare aircraft.

Photo 8.5   The MH-53E Sea Dragon is a helicopter designed for mine warfare

Photo 8.6   The C-2A Greyhound is a transport aircraft that shuttles supplies and personnel to and from aircraft carriers.

Trainer aircraft are generally two-seat fixed-wing or rotary-wing aircraft that allow instructors and students to go aloft together to learn or perfect the techniques of flying.

IDENTIFYING AIRCRAFT

There are thousands of aircraft in the Navy and identifying them can be a bit of a challenge. To accomplish this, they have names and official letter and number designations to distinguish one from the other. This designation system is explained in some detail below, followed by a simpler version for those who do not need or want quite so much precision. In fact, you can probably skip to that section ("The Simpler Way") if you do not deal with aircraft much and only need the basics, but I recommend you review the intervening sections first for a better understanding even if you eventually rely on the shorter version.

One thing to keep in mind (to avoid confusion) is that the current aircraft designation system has been in effect only since 1962, so if you are reading about aircraft in World War II, for example, the aircraft designations will not be the same.

Aircraft Names

Many types, designs, and modifications of aircraft form the naval air arm of the Navy. Like ships, aircraft have names, usually chosen by the designers or developers and approved by the Navy. The names are the most fun—there have been some great ones, like Phantom, Corsair, Hellcat, and Banshee—but they are the least revealing when it comes to type, mission, and so on. Unlike ships, individual aircraft do not have specific names but are identified by a system of "tail numbers" instead. More comparable to "class" in ships, all aircraft of a certain type bear the same name. For example, there are more Hornets in the Navy than you would care to count.

Aircraft Designations

More revealing than the names is a system of letters and numbers (sometimes referred to as the MDS [Mission Design Series] system) that is used to distinguish among the many types and variations of naval aircraft in service. The aircraft designation is a letter and number combination that tells you certain basic facts about the aircraft. The bad news is that this system is a little intimidating at first—when you first encounter "F/A-18E/F," representing the Super Hornet, you can't help but think this is something only for cryptanalysts—but the good news is that it *is* mostly logical and definitely decipherable. Some more good news is that this system of identification is the same for all the armed forces.

Be careful that you do not confuse formal aircraft designations with other abbreviations that sometimes come into play. For example, a concept that has been in development for some time is the Joint Strike Fighter, a fighter aircraft designed from the beginning to meet the various needs of all the armed forces, rather than just one, as has often been the practice. This aircraft is often referred to as the JSF, but this is *not* an aircraft designation within the formal system. Another example of this is "LAMPS," for "Light Airborne Multipurpose System"; though this actually is meant to describe the entire system (aircraft, platform, and associated equipment), you will often hear people refer to the LAMPS helicopter.

Probably the simplest way to begin deciphering these designations is to remember this: *one thing common to all aircraft designations is the dash.* Whether the aircraft is an S-3B, or an EA-6B, or an F/A-18 E/F, there is always a dash in the designation. If you use that as your starting point, you will have a consistent reference from which to begin cracking this code. Think of the dash in this system as being much like the decimal point in a number system. In mathematics, where numbers appear in relation to the decimal point indicates their value (tens, hundreds, positive, negative, etc.); in the aircraft designation system, where letters or numbers appear relative to the dash helps you understand their meaning.

In the forthcoming explanation, let's take a few examples from the real world by "decoding" the following aircraft designations: T-45A, SH-60B, EA-6B, NKC-135A, and F/A-18C/D.*

Type or Mission

Let's begin with the first letter to the left of the dash. It tells you one of two things: either the *type* or the basic *mission* of the aircraft.

---

*All these aircraft are in the Navy's inventory as of this writing, except NKC-135A Stratotanker, which is an Air Force plane. It is included because it is an example of an aircraft identified by three letters to the left of the dash. Currently, no Navy aircraft are so designated, but one could be in the future.

By "type," we mean whether it is a regular airplane (with fixed wings, engine(s), etc.) or some special kind of aircraft, like a helicopter. There is no letter for a regular airplane type (it is assumed by omission), but the following letters are used to tell you that the aircraft is a special type, as indicated:

G	glider
H	helicopter
Q	UAV (unmanned aerial vehicle)
S	spaceplane
V	V/STOL (vertical/short take off and landing)
Z	lighter than air (dirigible, etc.)

By "mission," we mean the primary purpose of the aircraft. The following letters describe aircraft missions as indicated:

A	attack
B	bomber
C	transport
E	special electronic installation
F	fighter
L	laser
O	observation
P	patrol
R	reconnaissance
S	antisubmarine
T	trainer
U	utility
X	research

Now comes the tricky part. If the aircraft is a special *type* (such as a glider or a helicopter), the first letter to the left of the dash will be one of those from the type list previously shown: G, H, Q, S, V, or Z. If it is *not* one of these special types (in other words, it's just a regular fixed-wing airplane), then there is *no letter* indicating type. It will be understood that no type letter indicates a regular fixed-wing airplane type. If that is the case, then the first letter to the left of the dash will be one from the *mission* list above: A, B, C, E, and so on.

Because the first letter to the left of the dash will indicate *either* the type or the mission of the aircraft, and the two lists do not overlap (except in one instance), you can combine the two lists into one and translate them appropriately (keeping in mind that the letter used is indicating either a type or a mission).*

In our examples from above, we know that the T-45A is a regular fixed-wing airplane with a primary mission of *training*; the SH-60B is a *helicopter*; the EA-6B is regular airplane with a primary mission of *attack*; and the NKC-135A is a regular airplane with *transport* as its primary mission. The F/A-18C/D is a bit of a special case: the slash between the F and the A indicates that this is a regular airplane that has *two* primary missions: it is both a *fighter* and an *attack* aircraft.

Modified Mission

Now, let's look to the second letter to the left of the dash. This letter (if there is one) is called the mission modifier. The following letters, when appearing in this position, have the meanings indicated:

A	attack
C	transport
D	director
E	special electronic installation
F	fighter
H	search and rescue/MEDEVAC (medical evacuation)
K	tanker
L	cold weather
M	multi-mission
O	observation
P	patrol
Q	drone (unmanned)
R	reconnaissance

---

*One minor problem is that this system uses the letter "S" for both type (spaceplane) and mission (antisubmarine). You are not likely to find this a problem—spaceplanes are rare and usually obvious in context—but it is worth mentioning here just in case.

S	antisubmarine
T	trainer
U	utility
V	staff
W	weather

These letters can be combined with *either* a type or a mission indicator in the first position to the left of the dash to tell you more about what an aircraft is used for. In our examples, the T-45A has no mission modifier, so it remains simply a trainer aircraft; the "S" of the SH-60B helicopter tells us that it is used for antisubmarine warfare; the EA-6B original attack mission has been modified by adding special electronics; and the NKC-135A is used as a tanker (for refueling other aircraft in the air). Because the two letters ("F" and "A") are separated by a slash in the F/A-18C/D, the F is not a mission modifier but is considered coequal with the A, so this an aircraft that is capable of carrying out both fighter and attack missions with equal capability.

To better understand this process, consider that the H-60 helicopter has been modified into several different versions in today's Navy. The SH-60 version is used for antisubmarine warfare; the HH-60 is a helicopter of the same basic design, but this version is used for search and rescue purposes; and there are other versions as well.

Sometimes a mission modifier results in a change of the name as well. An example of this is the EA-6B. Originally, there was (no longer in service) an A-6 Intruder. It was (as indicated by the "A" to the left of the dash) an attack aircraft. This original design was significantly modified (by adding some seats and a whole lot of complex electronic equipment) so that it could perform electronic warfare missions. It was redesignated the EA-6 and renamed Prowler.

Status Prefix

One more place (the third) to the left of the dash is sometimes (not very often) occupied by a letter called the status prefix. These letters are used for aircraft that are in a special status as follows:

G	grounded
J	special test (temporary)
N	special test (permanent)
X	experimental

V   prototype
Z   planning

The only one of our selected examples that has a letter in this position is the NKC-135A, and we can see from the list above that it is being used for a special test (permanently). You will not often see these letters used, but it is good to be aware of them should you encounter them.

Left-of-the-Dash Summary

Before moving to the right side of the dash, review table 8.1 for a summary of all items that *may* appear to the left of the dash in an aircraft designation.

Design

Now it is time to consider what is on the *right* side of the dash. This side is a bit easier because it does not require any memorization or any "crib notes" for translation.

You will recall that earlier we established that aircraft designations are sometimes referred to as the MDS (Mission Design Series) system. The letters to the left of the dash make up the "mission" part of that, and the letters and numbers to the right comprise the "design" and "series" parts.

Immediately to the right of the dash is the design number. All this number means is that this aircraft is a specific design of the particular type or mission. The first design of a patrol aircraft was designated "P-1" and when a whole new design of an aircraft for patrolling was accepted by the Defense Department, it was designated "P-2," and so on.

In our chosen examples, the T-45A is the forty-fifth design of a trainer aircraft that has been accepted by DOD; the SH-60B is the sixtieth helicopter design; the EA-6B is the sixth attack aircraft design; and the NKC-135A is the 135th transport design. The F/A-18C/D is once again a bit of an aberration; it is *not* the eighteenth fighter/attack design, but *is* the eighteenth fighter design accepted by DOD. (Reason does not always prevail.)

Series

Many times the basic design of an aircraft is modified in some way, so that it is no longer the same aircraft as originally designed, but it has not been changed enough to warrant calling it a whole new design. To indicate this significant modification (version), a series letter is appended to the design number. The

Table 8.1.　Left-of-the-Dash Summary

Status prefix	Modified mission	Basic mission	Vehicle type
G permanently grounded	A attack	A attack	G glider
J special test	C transport	B bomber	H helicopter
N special test (temporary)	D director	C transport	Q unmanned
X experimental	K tanker	F fighter	S spaceplane
Y prototype	L cold weather	L laser	V V/STOL
Z planning	M multi-mission	O observation	Z lighter-than-air
	E special electronics	P patrol	(If none of the
	F fighter	R reconnaissance	above letters appear,
	H search and rescue	S antisubmarine	it is a "regular"
	O observation	T trainer	fixed-wing aircraft.)
	P patrol	U utility	
	Q drone (unmanned)	X research	
	R reconnaissance		
	S antisubmarine		
	T trainer		
	U utility		
	V staff		
	W weather		

aircraft in its original design is considered to be "A" in the series. The first modification would be "B," the next would be "C," and so on. "I" and "O" are not used because they might be confused with one and zero.

So we now know that the T-45A and the NKC-135A are both original designs (as indicated by the "A" series indicator) and that the SH-60B and the EA-6B have each been modified once. The F/A-18C/D is, once again, a special case. There are actually two different versions of the Hornet in service; one has only one seat and is designated the "C" version, whereas the "D" version has two seats. Because both are in service you will often see them listed as "C/D" when referring to them generically.

The Simpler Way

All the above information is provided for those who want to know precisely what the aircraft designation system is telling them. But below I have compiled all the codes covering type, mission, and status into one list. In truth, if you use this list to decipher the information to the left of the dash without concerning yourself where these particular letters fall, you will know the essentials about the aircraft, and only rarely get confused.

Most of the time, the letters will be clear in context. For example, if you are given a picture of an aircraft that has the designation ZSH-7A and it has rotating blades on it, the chances are the "Z" indicates that it is in "planning" rather than being a "lighter-than-air" craft, and the "S" more than likely indicates "antisubmarine" rather than "spaceplane." So, use the letters below without worrying about where they fall to the left of the dash and you will, in most cases, know all you need to about an aircraft:

A	attack
B	bomber
C	transport
D	director
E	special electronic installation
F	fighter
G	glider *or* grounded
H	helicopter *or* search and rescue
J	special test (temporary)
K	tanker
L	laser *or* cold weather
M	multi-mission
N	special test (permanent)
O	observation
P	patrol
Q	drone (unmanned)
R	reconnaissance
S	antisubmarine *or* spaceplane
T	trainer
U	utility
V	V/STOL *or* staff

W	weather
X	research *or* experimental
Z	lighter than air *or* planning

AIRCRAFT ORGANIZATION

The basic organizational element for naval aircraft is the *squadron*. Some squadrons are carrier-based, spending part of their time on board aircraft carriers; others are land-based and, if their mission requires it, periodically deploy to other locations. Some squadrons are subdivided into detachments and are scattered to various ships or bases.

Though squadrons often have informal names that have more to do with morale than identification ("Black Knights," "Diamondbacks," etc.), they are officially identified by letter-number designations that, like ship hull numbers, tell something about their mission while giving them a unique identity. The first letter in a squadron designation is either a "V" or an "H." The V indicates fixed-wing aircraft and H is used for squadrons made up entirely of helicopters. If a squadron has *both* helicopters and fixed-wing aircraft, it is designated by a V. In the days when there was a third type of aircraft, the lighter-than-air (or dirigible) types, squadrons of those aircraft were designated by a Z.

The letter or letters following the V or H indicate the squadron's mission or missions. For example, a squadron whose primary purpose is training pilots to fly fixed-wing aircraft would be designated "VT." By adding a number, an individual squadron takes on a unique identity; for example, "VT-3." The numbers, in most cases, have some logic to them—such as even numbers indicating Atlantic Fleet squadrons and odd numbers designating Pacific Fleet—but movement and the periodic establishment and disestablishment of various squadrons has clouded some of the original intended logic.

Some of the squadron designations you might encounter are:

HC	helicopter combat support
HCS	helicopter combat support, special
HM	helicopter mine countermeasures
HS	helicopter antisubmarine
HSL	light helicopter antisubmarine
HT	helicopter training

VA	attack
VAQ	tactical electronic warfare
VAW	carrier airborne early warning
VC	fleet composite
VF	fighter
VFA	strike fighter
VFC	fighter composite
VP	patrol
VQ	reconnaissance/strategic communications
VR	fleet logistics support
VRC	carrier logistics support
VS	sea control (antisubmarine warfare, etc.)
VT	training
VX	test and evaluation

Aircraft squadrons are typically grouped into larger organizational units called "air wings." A carrier air wing (CVW) is usually made up of about eight squadrons, each serving different but integrated purposes. With these various squadrons on board, an aircraft carrier can carry out a wide variety of missions.

QUICKREFS

- Aircraft terminology:

  fuselage = body of an aircraft

  flaps = provide increased lift

  rotors = helicopter blades (actually rotating wings)

  tail rotor = small "propeller" that stabilizes a helicopter

  landing gear = wheels, skids, skis, or floats that support aircraft on the "ground"

  tail hook = a hook at the rear of a fixed-wing aircraft that allows it to land on an aircraft carrier

  powerplant = engine that provides the thrust and lift

  Mach = unit of speed equal to the speed of sound (Mach 2 = twice the speed of sound, etc.). One Mach is roughly equivalent to 650 mph.

> *supersonic* = faster than the speed of sound
> *subsonic* = slower than the speed of sound
> *hypersonic* = capable of flying five times the speed of sound or more
> *stealth* = virtual invisibility to radar

- Types of aircraft:

    > *fighter* = primarily engages other aircraft
    >
    > *attack* = primarily attacks enemy surface targets on the ground or on the sea
    >
    > *patrol* = used to locate enemy forces
    >
    > *antisubmarine* = finds and attacks enemy submarines
    >
    > *mine warfare* = lays mines in enemy waters or locates and destroys enemy mines
    >
    > *command and control* = uses sensors, communications equipment, and computers to coordinate various operations in the battle space
    >
    > *electronic warfare* = equipped to jam enemy radars and conduct other EW operations
    >
    > *transport* = carries cargo or personnel
    >
    > *trainer* = used to train student aviators

- Like ships, aircraft have names (like Phantom, Hornet, etc.) but they apply to types of aircraft, not individual aircraft.
- The aircraft identification system (sometimes called the MDS [Mission Design Series] system) uses a system of letters and numbers placed in a specific order to provide useful information about the aircraft. Where these letters and numbers appear in relation to the dash, tells you specific information about the aircraft as follows:

    —Status Prefix • Mission Modifier • Basic Mission • Vehicle Type • *DASH* • Design number • Series

    Left of the dash:

A	attack
B	bomber
C	transport
D	director
E	special electronic installation
F	fighter
G	glider *or* grounded

H	helicopter *or* search/rescue
J	special test (temporary)
K	tanker
L	laser *or* cold weather
M	multi-mission
N	special test (permanent)
O	observation
P	patrol
Q	drone (unmanned)
R	reconnaissance
S	antisubmarine *or* spaceplane
T	trainer
U	utility
V	V/STOL *or* staff
W	weather
X	research *or* experimental
Z	lighter than air *or* planning

Right of the dash:

—*Design number*: Sequentially assigned. The first fighter design would be an F-1, the second would be an F-2, and so on.

—*Series letter*: Sequentially assigned (A, B, C, etc.) to indicate significant modifications (versions) to the original design, with "A" used for the original design; "I" and "O" are not used.

- Naval aircraft are organized for operational and administrative purposes into *squadrons*.
- Squadrons are identified by letter and number combinations, such as VA-7 or HS-3. The first letter tells you whether the aircraft in the squadron are airplanes (V) or helicopters (H). Other letters are used to describe primary missions (such as "P" for patrol, "T" for training, etc.). Some squadron designations you are likely to encounter are:

HC	helicopter combat support
HM	helicopter mine countermeasures
HS	helicopter antisubmarine
HSL	light helicopter antisubmarine
HT	helicopter training

VA	attack
VAQ	tactical electronic warfare
VAW	carrier airborne early warning
VC	fleet composite
VF	fighter
VFA	strike fighter
VP	patrol
VQ	reconnaissance/strategic communications
VR	fleet logistics support
VRC	carrier logistics support
VS	sea control (antisubmarine warfare, etc.)
VT	training
VX	test and evaluation

- Squadrons are grouped together into larger units called *air wings*. A carrier air wing consists of about eight squadrons that together give the carrier a great deal of capability and flexibility.

CHAPTER NINE

Weapons

The Navy's overall mission is to maintain sufficient military capability to effectively deter a would-be enemy from using military power against the United States and its allies, to defend against any attacks that might occur, and to take offensive action against the enemy once hostilities have begun. Weapons are the mainstay of the military. Without them, the Navy could not carry out its combat missions or defend its ships, planes, bases, and personnel.

BASIC TERMINOLOGY

To understand the weapons used by the Navy, you should first be familiar with the following terms.

Ordnance. This generic term applies to the various components associated with a ship's or aircraft's firepower: guns, gun mounts, turrets, ammunition, guided missiles, rockets, and units that control and support these weapons. Do not make the common mistake of spelling it "ordinance," which is something altogether different.

Weapon system. When a number of ordnance components are integrated so as to find and track a target and deliver fire onto it, this is called a "weapon system." For example, a gun would be a weapon, but the gun plus the radars used to find and track the target and the ammunition-handling equipment used to load it would be a weapon system.

Gun. In its most basic form, a gun is a tube (barrel) closed at one end from which a projectile is propelled by the extremely rapid burning of a propellant (gunpowder).*

Propellant. The explosive substance (such as gunpowder) that is ignited to produce the energy to move an object (such as a bullet) toward an intended target. Rockets and missiles also have propellants, but they continue to burn through part or all of the firing, whereas a projectile receives all of its energy at the moment of initial ignition.

Projectile. An object that is "projected" through the air or water toward a target. Bullets and artillery rounds are forms of projectiles. So are rockets, missiles, and torpedoes, though these are less frequently referred to as projectiles than are bullets or artillery rounds.

Round. A unit of ammunition consisting of a projectile, a propellant, an igniting charge, and a primer. One shot.

Rocket. A weapon containing a propulsion section to propel the weapon through the air and an explosive section used to do damage to an enemy. A rocket is unable to change its direction of movement after it has been fired.

Missile. Originally called a "guided missile" this weapon is essentially a rocket (that is, it has a propulsion section and an explosive section) that also has a *guidance section* that allows its direction to be changed in mid-flight in order to better hit the target. Missiles can be aerodynamic (have wings and other components to allow them to be flown like an airplane—change altitude, course, etc.) or ballistic (assume a free-falling trajectory after an internally guided, self-powered ascent).

Torpedo. A self-propelled underwater weapon used against surface and underwater targets. Some torpedoes function like underwater rockets in that they cannot be controlled once they have been launched, whereas other more sophisticated versions can be guided, like an airborne missile, after they have been launched. Some torpedoes also have the ability to act autonomously; once they are launched, their own sonar and processors guide the torpedo to its target without further commands from the launch platform.

Mine. An underwater explosive weapon put into position by surface ships, submarines, or aircraft. A mine explodes only when a target comes into con-

---

*There is also a newer (substantially different) technology called a "rail gun" that uses electromagnetic energy rather than a chemical propellant to impart the necessary energy to the projectile. Rail guns are under development but are not currently in actual use.

tact with it or near enough to allow the mine's sensors (magnetic, acoustic, pressure, etc.) to detect the target's presence. Some versions may launch a torpedo when a target is detected.

Depth charge. Antisubmarine weapons fired or dropped by a ship or aircraft and set to explode either at a certain depth or in proximity to a submarine.

Bomb. Generally, an explosive weapon, other than a torpedo, mine, rocket, or missile, that is dropped from an aircraft. Bombs used in this context are free-falling (that is, they rely on gravity and the force imparted by the movement of the aircraft that is dropping them, and they have no other propulsion power to deliver them to the target). Bombs can be either "dumb" (unguided) or "smart" (with a guidance system to steer them to their target).

Warhead. That part of a missile that carries the explosive or some chemical package (for incendiaries, etc.).

Launcher. The device used to put a missile or rocket into the air. Older systems use an above-deck reloadable arm or box, but more modern systems use a vertical launch system (VLS) that uses tubes that are mounted in the ship's hull, expelling the missile through an open hatch in the deck.

Fire control. This is *not* the means for fighting fires (that is called "damage control" in the Navy), but it is the means used to control the firing of weapons. This can include the use of optics, radar, or laser beams and is the means for getting projectiles, missiles, and torpedoes onto a target.

Weapons Designations

You have probably noticed while reading other chapters in this book that the Navy is fond of designation systems (like the one for ship hull numbers that gives us such things as "CVN" and "DDG," and another for aircraft, the MDS system, which comes up with such things as "F/A-18E/F"). Though these systems can be bewildering at first, they really are useful for conveying a lot of information in a very compact way. Memorizing them is for the very brilliant or those with a lot of time on their hands. For the rest of us, being familiar with the basics—coupled with a willingness to refer to the various tables in this book—will go a long way toward making us smarter.

In this chapter you will encounter more of the same. In the section on missiles and rockets, you will encounter another system that makes sense out of such things as "AGM-88" and "RIM-2D." In the section on fire control, you will find yet another similar (but different) system for identifying electronics equipment within the Department of Defense.

To make things just a bit more confusing, you will also encounter some weapons (particularly bomb guidance units, warheads, and some launchers) that are designated using a U.S. Air Force system that looks similar but is altogether different from the Navy system used to designate rockets and missiles. And you will find that the Navy uses some Army designations when referring to some of its weapons (small arms in particular).

And there is yet another system that serves as kind of a catchall for those things not covered by the others. This is the Navy's "MARK and MOD" system.

MARK and MOD System

The Navy assigns MARK and MOD numbers to many types of equipment not covered by other designation systems. Bombs, torpedoes, guns, fire control systems (and the gun directors within them), rocket motors, missile launchers, warheads, and so on fall within this system, which has been around since World War II.

The system is less revealing in some ways than some of the other designation systems, but it is relatively simple. Beginning with a simple word description of the item (such as "torpedo" or "launcher"), it is then specified by a unique MARK number and followed by a MOD number. For example,

Torpedo, MARK 46 MOD 5

The MARK number is assigned sequentially and is the means of distinguishing one similar item from another. If the Navy accepts a modified version, that is indicated by the MOD number. The original version is designated MOD 0 and the first modified version is MOD 1.

According to MIL-STD-1661 (the governing instruction for this system—which you do not want to read unless you really have to), the *name* of the item is to *precede* the MARK number, and it is to be separated from the latter by a comma, as you see in the example above. (Talk about military "precision"!) But in reality you will more often see such unorthodox (but perfectly understandable) references as "the MARK 46 Torpedo." The instruction goes on to say that items with adjective descriptors, such as a "missile launcher," should be written in "reverse nomenclature" (with a separating comma) as in "launcher, missile." Sarcasm aside, there is a reason for this standardization. When these items are listed in inventory documents and the like, they will be grouped more logically when alphabetized. In other words, all the torpedoes will be listed together:

Torpedo, MARK 14 MOD 5
Torpedo, MARK 46 MOD 5
Torpedo, MARK 48 MOD 1

And all fire control systems will be grouped together:

Fire Control System, Gun, MARK 75 MOD 2
Fire Control System, Gun, MARK 86 MOD 1
Fire Control System, Guided Missile, MARK 13 MOD 1

Although the MOD part of the system is important in some contexts, it is often expendable in normal usage. So you will often see, for example, the MARK 46 MOD 5 Torpedo referred to as simply the "MARK 46 Torpedo."

The prefix "EX" is used instead of "MARK" for experimental items. If an EX item is adopted for operational use, it will use MARK from then on but will retain the originally assigned number. When an item is redesignated from EX to MARK, the MOD numbers are restarted from zero. For example, if an experimental item is designated EX 37, and the MOD numbers 0 through 3 were assigned during development, once the item is put into operational service, the new designation would be MARK 37 MOD 0. However, if more than one of the versions (modifications) is put into service, say EX 37 MOD 1 and EX 37 MOD 3, the new item designations would be MARK 37 MOD 0 and MARK 37 MOD 1. It *is* logical, if not pretty.

The governing instruction also calls for the words "MARK" and "MOD" to be written unabbreviated and in uppercase. However, the instruction allows some variations, and in actual practice, you may well see MARK as "MK" (or even "Mk") and you may also see "Mark" and "Mod" used. Therefore, the following variations are acceptable and you will see them used often:

Mark 46 Mod 4 Torpedo
MK 46 MOD 4 Torpedo
Mk 46 Mod 4 Torpedo

Dashes and other punctuations should not be used (and are not sanctioned by the governing instruction), but in truth you will see such things as these:

Mark-46 Mod 4 Torpedo
Mk.46 Mod 4 Torpedo

MISSILES AND ROCKETS

The Navy has a great many missiles and some rockets in its weapons inventory. The chief advantages of rockets and missiles over gun and bomb systems is their extended range, and missiles are, of course, more effective than rockets because of their increased accuracy. The major disadvantage of these weapons is their added cost. Both missiles and rockets can be fired from either ships (including submarines) or aircraft.

Rocket and Missile Components

Rockets have three major components—the airframe, the powerplant, and the warhead. As already explained, missiles have a fourth component—the guidance system.

The *airframe* is the body of the rocket or missile, which determines its flight characteristics and contains the other components. It must be light, because the other parts are generally heavy. Airframes are made of aluminum alloys, magnesium, and high-tensile (high-stress) steel. These metals can withstand extreme heat and pressure.

The *powerplant* is similar in function to the engines of an aircraft except that the aircraft's engines are reusable while the missile's propulsion unit is expended in its one flight. The powerplant propels the rocket or missile, usually at very high speeds to minimize its chance of being shot down before reaching its intended target. Some must be able to operate at very high altitudes where there is little or no atmosphere and therefore are required to carry both the fuel and an oxidizer in order to sustain combustion. Other, less expensive powerplants are air-breathing plants that carry only the fuel, but they cannot operate above about seventy thousand feet. Some rockets and missiles are equipped with additional boosters to extend the range.

The *warhead* is the part that does the damage. Its explosive may be conventional or nuclear or it may carry some chemical package to make smoke, fire, and so on.

Missile Guidance Systems

The guidance system in missiles constantly corrects the flight path until it intercepts the target. There are several different basic types of guidance systems: inertial, terrain following, homing, command, or beam riding. Many missiles use a combination of two of these systems—one guiding the missile through mid-course and the other used during the terminal stage. Components of these guidance systems will reside either in the missile itself or in the launch platform (ship, aircraft, etc.).

Inertial Guidance

This type of guidance uses a predetermined path programmed into a computer in the missile itself before launch. Missile speed and direction are checked constantly, and the on-board computer makes corrections to keep it on course.

Terrain Following Guidance

Missiles using this sophisticated type of guidance are preprogrammed with known terrain characteristics along the intended flight path that the missile can "recognize" and use to maintain or adjust its course and altitude. Such information can be obtained from satellite imagery or other forms of intelligence.

Homing Guidance

In this type of guidance, the missile picks up and tracks a target by radar, optical devices, or heat-seeking methods.

In an *active* homing system, the missile itself emits a signal that is reflected off the target and picked up by a receiver in the missile.

In a *semiactive* homing system, the signal comes from the launch platform rather than from the missile itself and the reflected signal from the target is then received by the missile, which uses the information received to correct its flight.

A *passive* homing system does not require either the missile or the firing ship or aircraft to emit a signal but uses the target's emissions to home on. For example, some passive homers use a target's own radar signals to home on; a heat-seeking missile can home in on the heat put out by the target's engines.

Command Guidance

After the missile is launched on an intercept course, two separate radar systems track the target and the missile. A computer (on the launch platform—

not the missile itself) evaluates how the missile is doing in relation to the target and transmits orders to the missile to change its track as necessary to ensure that it hits the target.

Beam-Riding Guidance

This is an older system in which the missile follows a radar beam (transmitted by the launch platform) to the target. A computer in the missile keeps it centered within the radar beam. Several missiles may ride the same beam simultaneously. If the missile wanders outside the beam, it will automatically destroy itself.

Missile and Rocket Designations

Navy rockets and missiles are often identified by a three-letter designation, followed by a number. For example, the Sparrow missile is known as an

AIM-7

The first letter tells you the *launch platform* (for example, "A" is launched from an aircraft, "R" is launched from a surface ship, "S" is launched from a submarine). Common letters used are:

A	air
B	multiple
C	coffin
F	individual
G	runway
H	silo stored
L	silo launched
M	mobile
P	soft pad
R	ship
S	underwater

The second letter tells you the weapon's *mission* (for example, "I" shoots down other aircraft or missiles, "G" attacks ships or land targets, and "U" is used to attack submarines). Common letters used are:

D   decoy
E   special electronics installation
G   surface attack
I   intercept aerial
Q   drone
T   training
U   underwater attack
W   weather

The third letter tells you the *vehicle type*.

L   launch vehicle
M   guided missile
N   probe
R   rocket

The number is a unique identifier for the rocket or missile. If there are subsequent major modifications (versions) to the basic design of a particular rocket or missile, they are indicated by a follow-on alphabetically sequential letter (A, B, C, and so on). For example, the third modification of the RIM-2 would be designated RIM-2C, the fourth would be RIM-2D, and so on.

So we know that the Sparrow AIM-7 from the example above is a missile that is launched from an aircraft and shoots down other aircraft or missiles. It is the seventh missile of its type, and there have been no other modifications.

The above explanation of the weapons designation system has been simplified to cover the designations you are most likely to encounter. However, like most government things, the system is a bit more complex than I have indicated above. Note that there is an optional prefix letter (indicating the status) that you might encounter if you are involved in weapons testing and the like and has more letters for the types of launch platforms, missions, and vehicle types. Common letters used are:

J   special test, temporary
N   special test, permanent
X   experimental
Y   prototype
Z   planning

So you could see something like "XRGM-1C" for an experimental (X) shipboard (R) surface attack (G) missile (M) that has been modified three times (C).

Missile and Rocket Categories

Missiles and rockets can be launched from aircraft, ships, and submarines and, depending upon their intended target, may be categorized as air-to-air, air-to-surface, surface-to-air, and so on. Some can be used against air and surface targets alike.

Air-to-Air

These missiles are carried by naval aircraft to shoot down enemy aircraft or missiles. Some examples of the current ones in use are listed below.

Sparrow. Designated the AIM-7, this highly maneuverable radar-guided (semi-active homing) missile can attack enemy aircraft from any direction in virtually all weather conditions and has a range of more than thirty nautical miles.

Sidewinder. The AIM-9 is an all-weather heat-seeking (passive homing) missile with a range of five to ten nautical miles depending upon conditions.

AMRAAM. The AIM-120 is a radar-guided sophisticated missile with a range of approximately thirty miles. This one does not have a name like "Sparrow" or "Sidewinder" but merely goes by the acronym "AMRAAM," which stands for "advanced medium-range air-to-air missile." Plans are for this missile to eventually be the Navy's sole medium/beyond visual range (M/BVR) missile.

Air-to-Surface (or Air-to-Ground)

These missiles are fired from aircraft and can be used against ships at sea and inland targets.

Maverick. The AGM-65 is an infrared-guided missile designed for day or night sea warfare (antiship) and land interdiction missions that will be around until the current supply runs out (estimated to be the year 2015).

HARM. The AGM-88 is named for its capabilities as a "high-speed anti-radiation missile." It homes on enemy radar-equipped air defense systems. The AGM-88E version is called the AARGM (advanced anti-radiation guided missile) and has been upgraded with a number of sophisticated technologies to make it a formidable weapon against enemy sensors.

Penguin. A Norwegian-built supersonic missile with a range of about twenty-five nautical miles, the AGM-119 is fired by the LAMPS (Light Airborne Multipurpose System) helicopters that are embarked on many Navy surface ships.

Surface-to-Air

Mounted on ships, these missiles are designed to shoot down incoming enemy aircraft and missiles. These weapons can be used in concert with or instead of friendly interceptor aircraft.

Standard. These surface-to-air missiles currently in use on Navy cruisers, destroyers, and frigates are grouped together in several variations of what are called the Standard (RIM-66) missiles. They replaced the older Tartar, Terrier, and Talos missiles.

Sea Sparrow. A modified version of the Sparrow air-to-air missile, this missile is carried by ships having no Standard missile capabilities. It has a range of about ten nautical miles and is designed to provide close-in protection when other means of antiair defense have been ineffective.

Rolling Airframe Missile (RAM). A small, lightweight, infrared homing intended primarily as a short-range missile designed to shoot down incoming cruise missiles. It got its name because it rolls in flight to provide stability.

Surface-to-Surface (Cruise) Missiles

These missiles can be fired from surface ships to strike other surface ships or land targets and are therefore surface-to-surface missiles, but because the ones in the U.S. Navy's inventory may also be fired from submarines or from aircraft to hit surface targets, they are more often generically referred to as "cruise missiles." More specifically defined as an unmanned, self-propelled guided vehicle that uses aerodynamic lift (wings) to extend its range, a cruise missile is essentially an unmanned aircraft that is relatively inexpensive (compared to manned aircraft) to maintain and operate.

Harpoon. Because Harpoons can be fired from virtually every combatant in the Navy (surface ships, submarines, and aircraft) the Harpoon is designated as the RGM-84 (surface ship-launched version), the UGM-84 (submarine-launched version), and the AGM-84 (aircraft-launched version). It has a range of more than seventy-five miles, and a special type called SLAM (for Standoff Land Attack Missile) is used to attack land targets.

Tomahawk. The BGM-109 can be used in several variations, including a TASM (Tomahawk antiship missile), a TLAM (Tomahawk land-attack missile), and a TLAM(N) (nuclear) version. These missiles vary in range from more than 250 nautical miles in the TASM version to more than 750 nautical miles and more than 1,200 nautical miles in the TLAM and TLAM(N) versions, respectively.

Fleet Ballistic Missiles

By definition, ballistic missiles do not rely upon aerodynamic surfaces to produce lift and consequently follow a ballistic (free-fall) trajectory when thrust is terminated. A ballistic missile has a relatively short flight time, and defenses against them are difficult.

Some ballistic missiles are relatively simple weapons (such as the Scud missile used by Saddam Hussein during the Persian Gulf War), whereas others are highly sophisticated types (such as the Trident missile described below).

All extremely long-range missiles are ballistic. Intermediate-range ballistic missiles (IRBM) can reach targets up to fifteen hundred nautical miles away, whereas the ICBM has a range of many thousands of miles. These weapons can be equipped with *multiple independently targetable reentry vehicles* (MIRV), which allow one missile to carry multiple warheads that can each be separately guided to a different target.

The Navy's fleet ballistic missiles are submarine-launched ICBMs with MIRV capability and were designed for strategic deterrence and attack. The earliest version was the Polaris, followed by the Poseidon. Today, U.S. fleet ballistic-missile submarines carry the *Trident* missile. The most advanced version, the Trident II (UGM-133), has a range of more than six thousand miles and is capable of carrying up to eight independent thermonuclear warheads.

Missile Launching Systems

Earlier missile systems had "dedicated" launchers—separate magazine-loaded launchers for each type of missile. This took up valuable space on board ship and increased topside weight. Later launchers handled more than one type of missile but still had to be individually loaded. The newest shipboard launching system is the *VLS* (vertical launch system), used in later *Ticonderoga-* and all *Arleigh Burke*–class ships. Missiles are stored in below-deck tubes from which the missiles can be directly launched. Any needed mix of missiles can be fired right from these tubes in quick succession without the delays involved in reloading topside launchers.

BOMBS

Although you will see the word "bomb" sometimes used in other contexts (such as "a suitcase bomb" or "a terrorist bombing"), for military purposes, bombs are normally considered to be ordnance dropped from aircraft that use gravity and the forces imparted by the aircraft's motion as the means of delivery (as opposed to rockets and missiles, which have their own means of propulsion).

There is an old joke that says, "Bombs are highly accurate and reliable weapons—they always hit the ground." Though this makes a good point about the nature of bombs as a weapon, its wisdom has been somewhat superceded by the advent of new technologies that include laser-guided bombs. These weapons—often called "smart bombs"—still use gravity and aircraft motion as the means of delivery, but these more sophisticated versions can be guided in flight and steered onto a target.

Bombs are generally classed as explosive, chemical, or practice. General-purpose (GP) bombs, weighing anywhere from one hundred to two thousand pounds, are explosive-type bombs and are generally used against unarmored ships or ground targets that can be damaged by blast or fragmentation. Semi-armor-piercing (SAP) bombs are used against targets that are sufficiently protected so as to require the bomb to have some penetration capability in order to be effective. Fragmentation bombs are usually smaller explosives dropped in clusters against troops and ground targets.

Chemical bombs contain specialized chemical agents that are used for a specific purpose, such as smoke bombs that spread heavy smoke over the target area in order to conceal movements of ships or troops, or incendiary bombs that produce intense fire when ignited and can be used against troops and ground targets.

Practice and drill bombs used in training may be loaded with sand or water but are inert (carry no explosive) and will cause no damage other than simple impact.

MARK 80 Series

Low-drag, general-purpose (LDGP) bombs are used in most bombing operations and are classed as the MARK 80 series:

MK 81	250 pounds
MK 82	500 pounds

MK 83   1,000 pounds
MK 84   2,000 pounds

The bodies of these bombs are aerodynamically designed and relatively light; approximately 45 percent of their total weight is actual explosives. Their fuzes may be either nose or tail mounted and may be fired mechanically or electrically. These bombs are not guided.

JDAM

The joint direct attack munition (JDAM) is a special tail kit that converts existing unguided, free-fall bombs into accurate, adverse weather, "smart" munitions. This tail kit includes an inertial navigational system and a global positioning system component for guidance control.

JDAM is a joint U.S. Air Force and U.S. Navy program, and the Navy has adopted the Air Force's designations for these weapons, so don't be confused when you see them designated using the abbreviation GBU (which stands for "Guided Bomb Unit"). The GBU-31 uses the MK 84 (2,000-pound bomb); the GBU-32 uses the MK 83 (1,000-pound bomb); and the GBU-38 uses the MK 82 (500-pound bomb).

LGB

Another joint Navy—Air Force program, laser-guided bombs (LGB) are highly accurate munitions fitted with a laser guidance unit fitted onto the nose that is used to guide the bomb onto the target. These guidance units are used with various bombs, such as the MK 83 1,000-pound bomb. The bomb-dropping aircraft (or some other source) illuminates the target with a laser beam and the bomb follows the reflection of that beam onto the target.

Like JDAM, these weapons are also designated using the Air Force system, so you will encounter such designations as GBU-10 (using the MK 84 bomb), GBU-12 (using the MK 82), and GBU-16 (using the MK 83).

NAVAL GUNS

Guns have been a major component of naval armament for centuries. Early guns were highly inaccurate, often very dangerous devices that had to be loaded from the front end and aimed simply by pointing at a visible enemy. Today's guns are much more powerful and accurate, far safer, and aimed and controlled by sophisticated electronic and hydraulic systems.

Early cannons had smooth bores (inside the barrel) and usually fired round shot. Modern guns have rifling in their barrels, which is a network of ridges (called "lands") and grooves shaped in a spiral that causes an elongated projectile to spin on its long axis (much as a well-thrown football) as it passes through the barrel. This increases the range and accuracy of the gun.

Guns are not as important to naval ships as they once were. Sophisticated missile systems, with their greater range and superior accuracy, have largely taken the place of the gun as the mainstay of naval armament. There is, however, still a need for naval guns. Certain missions are better performed by guns, and missiles tend to be much more expensive than guns. Also, the developing technology of the "rail gun" may give the naval gun a new life in the future.

U.S. Navy guns are classified by their inside barrel diameter (caliber) and by their barrel length. These two figures are expressed in a rather cryptic manner that may seem confusing at first, but it makes sense once you understand what it is telling you. The first figure in a Navy gun classification is the inside barrel diameter, expressed in inches or millimeters (mm). The second part follows a slash and, when it is multiplied by the first number, tells you the length of the gun's barrel. Thus, a 5-inch/54 gun would have an inside barrel diameter of five inches and a barrel length of 270 inches (5 x 54 = 270). In practice, the barrel length is often omitted, so you will commonly hear references to "the 5-inch gun" or "the 76mm gun" rather than the longer versions.

In years past, guns such as the 8-inch/55 and the 16-inch/50 were the main armament of large cruisers and battleships. Today, the most prevalent guns in the U.S. Navy are the MARK 45 5-inch/54 (on cruisers and destroyers), the MARK 75 76mm/62 (on frigates) (sometimes referred to as the MARK 75 3-*inch* gun, because 76mm equals 2.992 inches), and a specialized close-in weapons system (CIWS) known as the MARK 15 20mm/76 Phalanx system (mounted on many ships as a protection against incoming missile attacks). The Navy is replacing the 5-inch/54 with a longer-barreled 5-inch/62 gun that extends the range from thirteen nautical miles to more than twenty-one nautical miles when using conventional ammunition, and out to sixty nautical miles when firing ERGM (Extended Range Guided Munition) rounds.

Many Navy ships also carry saluting guns, which are used for ceremonial purposes and have no combat capability.

TORPEDOES

The torpedo is a self-propelled, explosive-carrying, underwater weapon. Early torpedoes were basically of the "point-and-shoot" variety, but modern versions have a guidance system that markedly increases the accuracy of the weapon.

A torpedo consists of a tail, afterbody, midsection, and head. The tail section includes the screws, fins, and control surfaces. The propulsion system is contained in the afterbody. The midsection houses batteries, compressed air, or liquid fuel. The head contains the explosive charge, fuze, and any sensing (acoustic, magnetic, etc.) devices.

Typical torpedo guidance systems are preset, wire-guided, or homing. Preset torpedoes follow a set course and depth after they are launched. Wire-guided torpedoes have a thin wire connecting the torpedo and the firing vessel, through which guidance signals can be transmitted to the torpedo to direct it to intercept the target. Homing torpedoes are active, passive, or a combination of active and passive. Active versions depend on the sensing signals generated and returned to the torpedo through a sonar device inside the torpedo. Passive types pick up telltale signals (such as noise or magnetic disturbances) to home in on. In the active/passive mode, the torpedo searches passively until a target is acquired, then active terminal guidance finishes the job of taking the torpedo to the target.

Surface ships launch torpedoes from tubes mounted topside or just inside the skin of the ship, or they propel them to the target area with a rocket called an ASROC. Submarines launch torpedoes through specially designed tubes, and aircraft deploy their torpedoes by parachute so as to reduce the impact when the weapon strikes the water.

Navy torpedoes are designated by the MARK system. Currently, the most common torpedoes in service are the MARK 46 (fired from surface ships and aircraft) and the MARK 48 (fired from submarines). The MARK 50 ALWT (advanced lightweight torpedo) is gradually replacing the MARK 46.

MINES

Naval mines are passive weapons that are planted under the water to await the passage of enemy vessels to explode and do damage. Their advantage is that they operate independently (that is, no personnel are required to operate them once they have been planted). Their chief disadvantage is that they are indiscriminate (they can damage friendly or neutral vessels as well as enemy ones if precautions are not taken). You might be confused a bit if you read naval history and see the word "torpedo" used. In earlier times, what we now call a mine was called a "torpedo." Today they are, of course, very different weapons.

Mines can be classified according to the method of actuation (firing), the method of planting, and their position in the water.

Mines may be actuated by contact or influence. A contact mine fires when a ship strikes it. Influence mines may be actuated by the underwater sound generated in a passing ship's current, by the ship's magnetic field, or by the mine's sensitivity to reduced water pressure caused by a passing ship.

Mines may be planted by surface craft, submarines, and aircraft. Planting mines using surface craft is the most dangerous method because the ship doing the planting is vulnerable to attack. Submarines can plant mines more secretly and aircraft are able to plant mines quickly and with less risk.

Moored contact mines are anchored in place and float near the surface of the water where a ship might strike them. Bottom mines, which lie on the ocean floor, are used only in relatively shallow water. They are influence mines, set off by sound, magnetism, or pressure.

FIRE CONTROL

A weapon, however powerful, is only as good as its accuracy. The process by which a projectile, missile, bomb, or torpedo is guided to its target is called fire control (or weapon control) and will, in each case, consist of the same basic phases:

- Target detection
- Target tracking
- Target evaluation
- Target engagement

A potential target is first detected by a sensor of some kind (such as radar or sonar). The target is then tracked and evaluated, either by human judgment, or by computer, or by a combination of the two. If the target is evaluated to be hostile, a decision is made, according to prescribed weapons doctrine, whether or not to engage. If the target is to be engaged, the appropriate weapon is selected. All available information is assimilated to produce a fire-control solution that will guide the weapon to the target. The weapon is then fired and the target destroyed.

Sensors

Before electronics arrived on the scene, enemies were detected and aimed at using the human senses, primarily the eyes. Modern weapons rely on electronic systems for the detection and tracking of targets. Most common are radar and sonar. Both operate on the same principle but differ in the medium used.

Radar

In its most elemental form, radar (originally an acronym for "radio detection and ranging") uses a transceiver to send out (transmit) a radio-like electronic signal that reflects off a target and then returns the signal to a receiver where a very accurate timing system measures the amount of time that the signal took to travel to and from the target and, using the known speed of the signal, calculates the range to the target. A built-in direction-finding system also provides a bearing (direction) to the target.

Radio waves work well in air (or in space), so radar is used to detect air and surface targets. Radio waves do not travel well under water, but sound does, so for subsurface targets, sonar is the sensor of choice.

Sonar

Sonar works on the same principle as radar, except that the signal used is sound rather than radio waves. Most of us have seen a movie or two in which the "pinging" sound of sonar makes for great dramatic effect. What is often overlooked is that sonar systems are more often used passively than actively (which means there would not be any pinging to hear).

The effectiveness of sonar is determined by a number of factors, most significantly by water pressure (a function of depth), temperature, and salinity. These conditions vary and have a direct effect on how a sonar system functions (range, sensitivity, etc.).

Sonar systems can be permanently mounted to the ship's hull or can be trailed along at a significant distance behind the ship on a long cable (called a "towed array" sonar); the latter allows the sonar sensors to get away from the sounds of the parent ship (particularly the sound created by the ship's propellers, which make hearing directly astern nearly impossible when using a hull-mounted sonar).

Sonar capability (both active and passive) can be extended well beyond the ship by having aircraft drop sonobuoys into the water. These disposable devices are smaller versions of a sonar system that float for a time and send sonar information back to the ship by transmitted radio signals.

Active versus Passive

As explained above, radar and sonar can be used to send signals (either radio or sonic) out to bounce off of a target and thereby gather information as to the target's range and bearing. Though this is the most effective method of targeting, it has a down side. If the targeted enemy has the right kind of receiving equipment, it can detect that signal. And what is worse, the enemy can pick up the signal at a greater distance than the sender, because the latter's signal must have enough power to go out and come back, whereas the targeted one only has to receive the transmitted signal one way.

A simple illustration of this is that sonar pinging sound mentioned above. Let's say that submarine A is trying to find submarine B and sends out a sonar "ping" to accomplish this. If sub A's sonar has just enough power for the sound to go out one mile, bounce off sub B, and return to sub A's sonar transceiver, then it has enough energy to actually travel *two* miles through the water (one mile out and one mile back). That means that sub B will actually be able to hear the ping at a range of two miles, so sub B will have the advantage of knowing that sub A is in the area before sub A knows sub B is present. With a direction-finding capability on the receiver, sub B will also be able to get a bearing on sub A; it will not, however, be able to get a range, because that relies on a signal being timed as it goes out and back.

The submarine sending out a ping is in "active" mode, whereas sub B (who is merely listening) is in "passive" mode. This same principle applies to radar. A ship equipped with the right kind of receivers can passively detect other ships' radars and know they are present without seeing them (because radar goes beyond the horizon) and without using its own radar to send out a signal. Radio transmissions used for communications are also vulnerable to this passive detection.

Electronic Warfare

The principles of active transmission and passive reception as used above, along with some additional measures (such as jamming or tampering with enemy signals), are considered to be forms of what is appropriately called electronic warfare (EW).

When a ship is taking advantage of these principles and is deliberately not "emitting" (sending out any signals)—that is, it is operating passively rather than actively—it is said to be in an "EMCON" (emission control) condition.

The gathering of electronic information passively is called "electronic warfare support measures" (ESM). Trying to "fool" the enemy by changing his signals in some way or jamming them is called "electronic countermeasures" (ECM). Trying to undo an enemy's countermeasures is called (you guessed it) "electronic counter-countermeasures" (ECCM). The latter is also sometimes referred to as "electronic protective measures" (EPM).

Identifying Electronic Equipment

Radar, sonar, and other Navy electronic equipments are identified by the *Joint Electronics Type Designation System* (JETDS). This system was developed during World War II and was originally called the "Army-Navy nomenclature system." Even though it is a joint system including the Air Force, Coast Guard, and Marines, it still retains the prefix identifier "AN" (for "Army-Navy"), and you will sometimes hear it called the "AN System."

The rest of the designation consists of three letters plus a number (with an optional letter sometimes following). Each letter tells you something about the equipment, and the number is the series number. As with other designation systems, different versions (modifications) of the same basic equipment are indicated by a sequential alphabetical letter; the original version has no letter, the first modification would be "A," the second "B," and so on. For example, you might encounter the JETDS designator

AN/SPY-1D

The "AN" simply tells you that this is a designation within JETDS. Because that is the only function of these letters, you will often see them dropped, so you might simply see "SPY-1D" referring to this same equipment. When the AN is used, it is followed by a slash (/) to avoid confusion.

The first letter ("S" in our example) indicates *where* this equipment is installed (the "platform" using it). These letters are used as follows:

A	piloted aircraft
B	underwater, mobile (submarine)

D	pilotless carrier (missile, drone, UAV)
F	fixed ground
G	general ground
K	amphibious
M	mobile ground
P	portable (by man)
S	surface ship
T	transportable (can be moved but not operated while in motion)
U	general utility
V	vehicle
W	water (surface/underwater combination)
Z	piloted/pilotless airborne combination

The second letter ("P" in our example) tells us the *type* of equipment. These letters are used as follows:

A	invisible light, heat radiation (i.e., infrared)
B	COMSEC (secure communications)
C	carrier (electronic wave or signal)
D	RADIAC (radioactivity detection, identification, and computation)
E	laser
F	fiber optics
G	telegraph or teletype
I	interphone and public address
J	electromechanical
K	telemetering
L	countermeasures
M	meteorological
N	sound in air
P	radar
Q	sonar and underwater sound
R	radio
S	special or combination
T	telephone (wire)
V	visual/visible light
W	armament (only used if no other letter applies)

X   fax or television
Y   data processing
Z   communications

The third letter ("Y" in our example) defines the *purpose* of the equipment. These letters are used as follows:

A   auxiliary assembly
B   bombing
C   communications
D   direction finding, reconnaissance, and surveillance
E   ejection or release
G   fire control or searchlight directing
H   recording or reproducing
K   computing
M   maintenance or test
N   navigation aid
Q   special or combination
R   receiving or passive detecting
S   search and detecting (range and bearing)
T   transmitting
W   automatic flight or remote control
X   identification or recognition
Y   surveillance (target detecting and tracking) and control (fire control and air control)
Z   secure

From the above lists, we can determine that our example (AN/SPY-1D) is a surveillance and control (Y) radar (P) that is installed on surface ships (S).

The number after the letters ("1" in our example) is the *model number.* Each new model of an equipment is assigned the next number in sequence, so if the Navy acquired a new surveillance and control radar for use on surface ships, it would be designated "SPY-2."

As with most of these designation systems, this one also uses an appended letter at the end to tell you that this is a modified version from the original design, with some differences. The first (original) version has no suffix. The first modification uses "A," the second would be "B," and so on. The letters

"I," "O," "Q," "S," "T," "X," "Y," and "Z" are not used for reasons too complicated to go into here.

With what we now know about electronic designations, we can determine that an

AN/PRC-66B

is a portable (P) radio (R) used for communications (C) purposes with a model number of "66," and this is the second modification (B) to the original design.

Weapon-Control Systems

Ships, aircraft, and submarines all incorporate various types of weapon-control systems. Surface- and air-search radars have been continuously improved since World War II to detect high-performance targets at long ranges in any weather. The newer surface-ship control systems work with guns and missiles, and they include radars and digital computers that can quickly acquire and track targets while directing shipboard weapons.

There are various systems in use that are beyond the scope of this book for a detailed look, but one system you are likely to encounter is the Aegis system, a rapid-reaction, long-range, fleet air-defense system capable of effectively handling multiple surface and air targets simultaneously. It includes the very capable AN/SPY-1 radar, a quick-reaction tactical computer for overall command control, a digital weapon-control system, and state-of-the-art guided-missile launchers. Found in *Ticonderoga*-class cruisers and *Arleigh Burke*—class guided-missile destroyers, the Aegis system gives a force commander the capability of controlling all the surface and aerial weapons of an entire battle group in a multithreat environment.

Even though it is an incorrect usage, you will hear the expressions "Aegis-class cruiser" (describing the *Ticonderoga*-class CGs) and "Aegis-class destroyer (referring to the *Arleigh Burke*—class DDGs). Some habits are impossible to break.

Submarines and aircraft have their own control systems, similar in general principle to those used in surface ships.

Fleet ballistic missiles fired from submarines are controlled by a missile fire-control system, which is connected to the submarine's inertial navigation system. This highly sophisticated electronic navigation system keeps accurate track of the ship's position. When missiles are to be fired, the fire-control system takes current position data and quickly computes firing information to put missiles on the proper ballistic course. While in flight, the missile keeps itself on course with the aid of a built-in navigational system.

SMALL ARMS

The Navy also uses a variety of small arms (pistols, rifles, shotguns, grenade launchers, and machine guns) for various purposes, including sentry duty, riot control, and landing parties. These weapons have become much more prominent since the attack on USS *Cole* in Yemen in October 2000. Since then, you are more likely to see large machine guns mounted on ships in port, for example.

A detailed discussion of these weapons is beyond the scope of this book, but some basic facts are worth noting.

Classifying Small Arms

Just as with larger Navy guns, small arms are often differentiated by the inside diameter (bore) of the barrel. Like larger naval guns, this diameter may be expressed in either inches or millimeters, but unlike larger guns, small arms do not include a follow-on figure representing the length of the barrel (see the discussion of naval guns above). When the figure is in inches, it is referred to as "caliber," as in ".45-caliber pistol," but when it is expressed in millimeters, the term "caliber" is not used, as in "9mm pistol."

Shotguns are an exception. They are usually differentiated by "gauge."* This also refers to the bore but is defined as the number of lead balls of that particular diameter required to make a pound. For example, it would take twelve lead balls of the diameter of the 12 gauge shotgun to equal one pound and sixteen balls for the 16 gauge shotgun. This means that the 12 gauge shotgun has a larger bore than the 16 gauge, which seems backwards at first but makes sense when you think about it.

Any weapon with a bore diameter of 0.6 inches (.60-caliber) or less is called a small arm. The largest Navy small arm is the .50-caliber machine gun.

Small arms are considered to be "automatic" if holding down the trigger causes the weapon to continuously fire and "semiautomatic" if the weapon

---

*Note: Sometimes you will see "gage," but the preferred Navy spelling is "gauge."

reloads automatically when fired but requires another pull of the trigger to fire off another round.

Some small arms you may encounter are identified by the Army system of terminology. An "M" preceding a number identifies a particular weapon, such as the "M14 rifle." Modifications are identified by a follow-on letter and number combination. For example, the M16 rifle was modified twice as the M16A1 and M16A2 versions. Sometimes the Navy system of "MARK" and "MOD" (discussed earlier in this chapter) is used, as in the "20mm MK 16 MOD 5 machine gun."

Pistols

One of the oldest weapons in the Navy inventory is the M1911A1 .45-caliber semiautomatic pistol. It is commonly (though erroneously) referred to as the "45 automatic." Because you must pull the trigger each time you fire a round, this pistol is actually *semi*automatic. Its magazine (a removable ammunition holder that feeds rounds into the chamber) holds seven rounds and it has a maximum range of a little more than sixteen hundred yards but is usually effective only at about fifty yards. One of its chief advantages is stopping power—whereas a .38-caliber revolver can be just as lethal as the .45, the latter is more likely to knock a man off his feet, even one who is pumped up on adrenalin. This can be a major asset when dealing with a charging fanatic, for example.

The 9mm M9 semiautomatic pistol is a similar weapon to the .45-caliber pistol. Slightly lighter in weight, it has a maximum range of eighteen hundred meters (1,962.2 yards) and an effective range of fifty meters (54.7 yards). A major advantage is that its magazine has a capacity of fifteen rounds, more than double that of the .45 pistol.

The .38-caliber revolver has maximum and effective ranges similar to the .45 pistol but is lighter in weight. This makes it more suitable for flight personnel. It has a six-round capacity and its relatively simple design makes it unlikely to jam.

The 9mm pistol is the official replacement for both the .45 and .38 pistols, but you may find the latter weapons still in service at some commands.

Rifles

In the Navy, you may encounter the M16 rifle, which is a magazine-fed weapon that fires a 5.56mm (just slightly larger than a .22-caliber) round. A major advantage of this weapon is its light weight (7.5 pounds on earlier models and 8.5 pounds on later ones); this is a significant reduction from its predecessor, the M14, which weighed in at eleven pounds. The 5.56mm caliber may seem small, but the high muzzle velocity (more than three thousand feet per second) makes this a very powerful weapon. The M16A1 has a selector lever that allows the user to fire in automatic or semiautomatic mode, and the M16A2 has a similar selector that permits semiautomatic or burst (three rounds) modes. The magazine capacity is either twenty or thirty rounds, depending upon the type used, and the maximum range is 460 meters (503 yards).

An option to the M16 rifle that is sometimes preferred by Special Forces personnel is the M4 carbine, a shorter variant of the former. The M4 carbine is sometimes used when the firepower of the M16 is desirable but the weight of the M16 is not.

Though the M16 is the replacement rifle for the Navy, you may still encounter some M14 rifles. Firing a 7.62mm round in either automatic or semiautomatic mode, the M14 was the last of the wooden-stock rifles before lighter, plastic ones appeared on the M16.

Shotguns

The most common shotgun in the Navy is the Remington M870. Manually operated, this 12 gauge, pump-action shotgun can fire four rounds without reloading. The Mossberg M500 is another 12 gauge shotgun similar to the M870 that is sometimes used.

Machine Guns

The .50-caliber M2 Browning machine gun (abbreviated "BMG") is mounted on many surface ships and patrol craft for close-in defense. Ammunition is belt-fed at a rate of 450 to 500 rounds per minute. The BMG has a maximum range of seventy-four hundred yards and an effective range of two hundred yards.

A lighter, but very effective, machine gun is the 7.62mm M60. With a maximum range of 3,725 meters (4,075 yards) and an effective range of 1,100 meters (1,200 yards), the M60 was originally designed for use by ground troops but has been adapted for naval use as well. Although Stanley Kubrick and other Hollywood directors have different ideas, firing in short bursts is preferable to continuous firing to prevent overheating.

Grenade Launchers

You may encounter three different kinds of grenade launchers in the Navy. The 40mm M79 is handheld, like a shotgun, and fires one round (grenade) at a time. The MK 19 MOD 3 machine gun is a mounted weapon that fires multiple 40mm grenades in fully automatic bursts. The M203 grenade launcher is actually an accessory that can be attached to the M16 rifle.

QUICKREFS

- Naval weapons include the following types:
  —*Gun*. Imparts energy to (fire) a projectile (bullet) by rapidly burning a propellant in its barrel.
  —*Rocket*. Has self-contained propellant that continues to burn in flight.
  —*Missile*. A form of rocket that can be controlled (guided) in flight.
  —*Torpedo*. A kind of underwater missile.
  —*Mine*. Placed underwater (non-moving) and detonated by passing ships (by contact, sound, etc.).
  —*Bomb*. An explosive dropped from an aircraft that relies upon gravity and the motion of the delivering aircraft to get it on target; can be unguided or guided ("smart").

- The common forms of *missile guidance* are inertial, active homing, semi-active homing, passive homing, command, and beam-riding.
- Missiles and rockets can be *air-to-air* (launched from an aircraft to hit another aircraft or missile), *air-to-ground* (launched from an aircraft to strike a land target or a vessel at sea), *surface-to-air* (launched from a surface ship or land installation to strike an aircraft or missile in the air), or *surface-to surface* (launched from a ship or land installation to strike a vessel or land target).
- *Cruise missiles* are a form of surface-to-surface missile that can also be launched from aircraft and submarines. These missiles have aerodynamic wings to sustain them through flight and are therefore very maneuverable and long-ranged.
- *Ballistic missiles* do not rely on aerodynamic surfaces for lift and follow a free-fall trajectory after thrust is terminated. The Navy's fleet ballistic missile is the *Trident*, which is an *ICBM* (intercontinental ballistic missile) equipped with *MIRV* (multiple independently targetable reentry vehicles) warhead technology.

- The Navy still uses conventional *LDGP* (low-drag, general purpose) bombs in the MARK 80 series but increasingly employs more sophisticated ("smart") bombs called *JDAM* (joint direct attack munition) and *LGB* (laser-guided bomb). LDGPs are simple bombs in various sizes (500 to 2,000 pounds) that are simply dropped on targets, whereas JDAMs and LGPs are these same bombs with special attachments that use inertial and GPS navigational information or laser beams to guide them onto the target.
- Naval guns are less expensive than missiles and are better suited for certain missions (like support for amphibious landings). The most common naval guns today are the *MARK 45 5-inch/54* (five inches is the inside barrel diameter and that number is multiplied fifty-four times to get the barrel's length) on destroyers and cruisers, the *MARK 75 76mm/62* on frigates, and the *MARK 15 20mm/76* close-in weapons system found on many Navy ships.
- Modern torpedoes are essentially underwater missiles using preset, wire-guided, or homing systems for guidance. They are launched from surface ships or aircraft (*MARK 46*) or from submarines (*MARK 48*). The MARK 46 is gradually being replaced by the *MARK 50 ALWT* (advanced light-weight torpedo).
- Naval mines can be planted by ships, submarines, or aircraft and are activated (detonated) by contact, magnetic field, pressure change, and so on.
- Weapon's control systems rely on various sensors (radar, sonar, and electronic warfare). These can be used actively (where the weapon platform transmits and receives signals) or passively (where the platform only receives).
- Forms of EW (electronic warfare) include:
  —*EMCON (Emission Control).* Limiting usage or completely shutting off one's own electronic emitters (such as radio, radar, and sonar) to avoid enemy detection.
  —*ESM (Electronic Warfare Support Measures).* Detecting enemy electromagnetic emissions.
  —*ECM (Electronic Countermeasures).* Actively countering (such as jamming) an enemy's electromagnetic emissions.
  —*ECCM (Electronic Counter-Countermeasures).* Undoing or protecting against an enemy's ECM against you. Also called "EPM" (Electronic Protection Measures).

- *Ticonderoga*-class cruisers and *Arleigh Burke*—class destroyers employ a highly sophisticated, rapid-reaction, long-range, fleet air-defense system capable of effectively handling multiple surface and air targets simultaneously known as the *Aegis* system. This system is coupled with the *VLS* (vertical launch system) that allows different kinds of missiles to be stored in, and directly launched from, tubes mounted in the ships' hulls.

- The Navy uses small arms (pistols, rifles, shotguns, etc.) as well as its larger weapon systems.

- Unfortunately, there are several different systems used for identifying weapon system components. There are the MARK and MOD system, the Missile and Rocket Designation system, and the Joint Electronics Type Designation System (JETDS). To further complicate things, some weapons components are identified by the designation systems of other services.

- Many weapons components—such as bombs, torpedoes, guns, and missile launchers—fall within the *MARK and MOD* designation system.

  —The designation begins with a word description, such as "torpedo," and is followed by a unique MARK number and then a MOD (modification) number (as in "Torpedo, MARK 46 MOD 5").

  —If the descriptive word is enhanced by an adjective (as in "guided missile"), the adjective follows the main word, separated by a comma (as in "missile, guided").

  —In normal usage, you will often see forms other than the one described above, so that the "Torpedo, MARK 46 MOD 5" becomes any of the following:

  MARK 46 MOD 5 torpedo
  MARK 46 torpedo
  Mark 46 Mod 5 torpedo
  MK 46 torpedo
  Mk 46 Mod 5 torpedo

- The basic *Missile and Rocket Designation* system consists of three letters followed by a number (as in "AIM-7"). See Table 9.1

  The first letter tells you the *launch platform.*
  The second letter tells you the *mission.*
  The third letter tells you the *type.*
  An optional prefix letter can be used to tell you of any *special status* (such as experimental).

Table 9.1   Missile and Rocket Designation System

Status (optional)	Launch platform	Mission	Type
J special test, temporary	A air	D decoy	L launch vehicle
N special test, permanent	B multiple	E special electronics installation	M guided missile
X experimental	C coffin	G surface attack	N probe
Y prototype	F individual	I intercept aerial	R rocket
Z planning	G runway	Q drone	
	H silo stored	T training	
	L silo launched	U underwater attack	
	M mobile	W weather	
	P soft pad		
	R ship		
	S underwater		

An optional sequential letter (A, B, C, etc.) can follow the number to tell you the *modification* (version).

- *Joint Electronic Type Designation System* (JETDS) is another shorthand system that identifies military electronic equipment. A typical identifier would look like this: "AN/SLQ-32B." See table 9.2.

The first two letters (often omitted) followed by a slant (/) are always "AN" and identify the designation as part of the JETDS system.

The first letter after the slant indicates the *platform* (where the equipment is installed).

The second letter after the slant indicates the *type* of equipment.

The third letter after the slant defines the *purpose* of the equipment.

The numbers after the dash are the *model number*, and the last letter identifies the *version*.

Table 9.2 Joint Electronic Type Designation System

Installation		Type of equipment		Purpose	
A	piloted aircraft	A	invisible light, heat radiation (i.e., infrared)	A	auxiliary assembly
B	underwater, mobile (submarine)	B	COMSEC (secure communications)	B	bombing
D	pilotless carrier (missile, drone, UAV)	C	carrier (electronic wave or signal)	C	communications
F	fixed ground	D	RADIAC (radioactivity detection, identification, and computation)	D	direction finding, reconnaissance, and surveillance
G	general ground	E	laser	E	ejection or release
K	amphibious	F	fiber optics	G	fire control or searchlight directing
M	mobile ground	G	telegraph or teletype	H	recording or reproducing
P	portable (by man)	I	interphone and public address	K	computing
S	surface ship	J	electromechanical	M	maintenance or test
T	transportable (cannot be operated while in motion)	K	telemetering	N	navigation aid
U	general utility	L	countermeasures	Q	special or combination
V	vehicle	M	meteorological	R	receiving or passive detecting
W	water (surface/ underwater combination)	N	sound in air	S	search and detecting (range and bearing)
Z	piloted/pilotless airborne combination	P	radar	T	transmitting
		Q	sonar and underwater sound	W	automatic flight or remote control
		R	radio	X	identification or recognition
		S	special or combination	Y	surveillance and control
		T	telephone (wire)	Z	secure
		V	visual/visible light		
		W	armament (only used if no other letter applies)		
		X	fax or television		
		Y	data processing		
		Z	communications		

CHAPTER TEN

Lucky Bag

Maintaining good order, discipline, and cleanliness aboard ship has been a high priority in the U.S. Navy from its earliest days. One method for achieving these things was the tradition of the "Lucky Bag." As the tradition goes, any personal items left out in the berthing compartment ("gear adrift") were confiscated by the Master-at-Arms and placed in a special bag. These items were later auctioned off—the funds used for the general welfare of the crew—thereby making "lucky" those Sailors fortunate enough to obtain new items for relatively little money.

This practice, of course, led to a varied assortment of unrelated items in the bag. In this chapter, we will review a number of miscellaneous topics that may be useful to you but do not rate a chapter of their own nor fit into any of the previous chapters.

STRANGE WAYS OF DOING THINGS

In dealing with Sailors, you will often hear them talking in strange ways, or writing dates in an odd manner, or telling time oddly. There are good reasons behind these practices, and though you might argue that they are not all necessary when individuals are away from their ships and aircraft, old habits are not easily altered, especially when most of those individuals are going back to sea at some point. It will help if you learn how to do things in these unconventional ways and it will probably save you come unnecessary confusion.

Numbers and Letters

When talking on a radio circuit, on a sound-powered telephone aboard ship, and even on a standard telephone, your voice does not have the same clarity that it does when speaking face to face with someone. That is one of the reasons why the Navy insists upon standardized terminology when communicating and has devised some unusual ways to avoid confusion.

Numbers

To ensure the clarity desired, numbers are often spoken individually rather than in the more conventional (but confusing) manner you are used to in civilian life ("one, two, zero" vs. "one hundred twenty"). Because the numbers "five" and "nine" sound very much alike on a sound-powered circuit or on a radio, the number "nine" is nearly always pronounced "niner" in the Navy. This will seem strange to you at first, but that strangeness goes away after a time, and you are not likely to confuse a five with a niner if you hear it used on a telephone, which is obviously a good thing.

Phonetic Alphabet

One of the most valuable tools in maintaining clarity and avoiding confusion in communications is the phonetic alphabet. C, D, E, P, V, T, and Z all sound very much alike on a telephone, for example. By using phonetic equivalents (Charlie, Delta, Echo, Papa, Victor, Tango, and Zulu) there is no chance of someone misunderstanding what letter is meant. If someone says, "We need part number six Alfa," instead of "six A," there is no chance that someone will confuse that with part 6K (which would be "six Kilo"). The phonetic alphabet can also be used to spell out unfamiliar words that someone is having difficulty understanding: "Echo-Xray-Echo-Golf-Echo-Sierra-India-Sierra" works a lot better on the telephone than "ee-ex-ee-gee-ee-ess-eye-ess"; unfortunately, the phonetic alphabet will *not* tell you what that word means!

The phonetic alphabet has been around for a long time, but has not always been the same. Back in the days of World War II, the phonetic alphabet began with the letters "Able, Baker, Charlie," K was "King," and S was "Sugar." After the war, when the NATO alliance was formed, the phonetic alphabet was changed to make it easier for the people who speak the different languages found in the alliance. That version has remained the same, and today the phonetic alphabet begins with "Alfa, Bravo, Charlie," K is now "Kilo," and S is "Sierra."

As you look over the phonetic alphabet, you will notice certain idiosyncrasies that you must accept. For example, the letter "Alfa" is spelled with an F instead of "ph." This is because some of our allies do not have a "ph" in their language. Also note that "Whiskey" ends in "key," not "ky," and "Juliett" ends in two Ts.

Note the pronunciations in the table provided, each word is accented on
the syllable in capital letters. This is no time for individuality—it is essential
that everyone say these words as much the same as is possible to avoid any
confusion, which is the whole purpose. The letter P is pronounced *pah-PAH*,
not *POP-ah* as most Americans are more likely to say. L is *LEE-mah*, not *LYE-
mah*, and Q is *kay-BECK*, not *QUAH-beck*.

a	Alfa	*AL-fah*
b	Bravo	*BRAH-vo*
c	Charlie	*CHAR-lee*
d	Delta	*DELL-tah*
e	Echo	*ECK-oh*
f	Foxtrot	*FOKS-traht*
g	Golf	*GOLF*
h	Hotel	*hoh-TELL*
i	India	*IN-dee-ah*
j	Juliett	*JEW-lee-ett*
k	Kilo	*KEY-loh*
l	Lima	*LEE-mah*
m	Mike	*MIKE*
n	November	*no-VEM-ber*
o	Oscar	*OSS-ker*
p	Papa	*pah-PAH*
q	Quebec	*kay-BECK*
r	Romeo	*ROW-me-oh*
s	Sierra	*see-AIR-rah*
t	Tango	*TANG-go*
u	Uniform	*YOU-nee-form*
v	Victor	*VIK-ter*
w	Whiskey	*WISS-key*
x	Xray	*ECKS-ray*
y	Yankee	*YANG-key*
z	Zulu	*ZOO-loo*

Radio Talk

Because many Sailors have to talk on radios as part of their profession, you
may note that some of that radio jargon appears in normal conversation.

Don't be surprised is someone says "affirmative" and "negative" instead of "yes" or "no." You will also hear "Roger" (or "Roger that") as a means of acknowledgement. You may well begin to emulate this kind of talk, which is fine, as long as you do it correctly. Hollywood loves to imitate military talk but absolutely refuses to get it right most of the time and thereby instantly loses credibility among military viewers. A few quick definitions (in layman's terms) will avoid much embarrassment.

- "Roger" means, "I heard what you said."
- Wilco" means, "I heard and understood what you said and I will do what you say" (*will co*mply).

Because the meaning of "Roger" is included in the meaning of "Wilco," only the uninformed and a Hollywood star would say "Roger, Wilco" together.

- "Over" means, "I am finished talking and now it's your turn to talk."
- "Out" means, "I am finished talking and no response from you is necessary (in other words, "This conversation is over").

Because "over" says one thing and "out" says something entirely different, you should *never* hear these two words used together (yet you often do). To say "over and out" is to say "I am finished talking and now it's your turn to talk and I am finished talking and you should say nothing."

Sad to say, there are times when someone who wants to sound military will say "roger, wilco, over, and out." Like nails on a chalkboard.

Bravo Zulu

This is not a hybrid version of two famous movies. It is the Navy's unique way of praising someone, of saying, "Well done." Remembering the phonetic alphabet explained above, you might be tempted to ask, "Why not 'Whiskey Delta?'"

The origins of the term can be found in the practice of using codes to convey information or orders at sea. Before radio was invented, ships needed some means of communicating, so a set of signal flags representing letters and numbers was created so that messages could be created and hoisted by one ship and read by other ships within visual range. But trying to spell out a message of very many words would require a lot of halyard space, so signal books were devised with simple codes that assigned longer meanings to short combinations of letters. For example, if all ships in a squadron had the same signal book that assigned the code letters "DCV" to mean, "Engage the nearest enemy from windward," the squadron commodore could order that for all his ships at once, simply by hoisting those three flags. This had the added advantage that, without the same signal book, the enemy would not know what the commodore was intending.

When radio was invented, the same codes could still be used to convey longer messages in shorter terms (over longer distances), simply by broadcasting the code letters by Morse code or voice. As explained above, for voice radio a phonetic alphabet was created to prevent a commodore intending to tell his ships to "engage the enemy" (DCV) from accidentally telling them to "return to port" (TCE). If they heard the letters wrong, such a mistake could hardly be career-enhancing.

The signal books, assigning specific meanings to different combinations of letters and numbers, have changed at times to keep enemy forces from becoming familiar with the codes and to reflect changes of technology. The codes also grew, incorporating administrative as well as tactical information. One of the signals included along the way was one that allowed a commander to send his approval for a successful evolution, to say, "Well done." During World War II, "Tare Victor George" was listed in the codebook with that meaning.

When a new codebook, called the Allied Naval Signal Book (ACP 175), was adopted after the creation of NATO in 1949, "Tare Victor George" was replaced by "Baker Zebra" for no other reason than randomness. It remained that way until 1957 when the new phonetic alphabet changed it to "Bravo Zulu."

One can presume that the Navy was doing a lot of things well through the years, because Sailors became used to (and coveted) the words "Bravo Zulu." It became so familiar and so traditional that subsequent changes to the signal book retained that one code when everything else changed. It remains today; so if someone says "Bravo Zulu" to you, you should smile and pat yourself on the back for a job *Well Done*.

Time

One can argue about the relative merits of many of the strange habits one encounters in the Navy, but nowhere does the military trump the civilian world more than in the telling of time. The "am" and "pm" system that we grew up using is familiar and therefore comfortable, but *dumb*! The twenty-four-hour system makes so much more sense because it does not lend itself to confusion. Note that it is not only used by all the military services but also by police forces, in hospitals, by air-traffic controllers, and in every other profession where the telling of time is important.

This system may be unfamiliar at first, but it is not difficult to learn. Very simply, instead of starting the twelve-hour cycle over again at noon, the twenty-four-hour system just keeps going, so that instead of 1:00 pm, we say 1300 (pronounced "thirteen hundred"). Instead of 2:00 pm, the time is 1400, and so on, until we reach midnight. How you get used to this is up to you. Some people remember it by subtracting two (or two hundred) from the strange times to get to their more familiar "pm" time, so that 1500 minus two (hundred) becomes 3:00 pm. Others just think in terms of a twenty-four clock and the times are rather logical. Whatever system you use at first, eventually it will become second nature, and unless you are very stubborn, you will come to appreciate the logic and simplicity of it.

Times other than the actual hour are virtually the same as in the twelve-hour system, so that 1:05 pm becomes 1305, 11:47 pm becomes 2347, and so on. Note that there are no colons used as there are in civilian time.

One other thing you must get used to is that zeros are used so that time is always expressed in four digits. This means that the "am" hours are virtually the same in both systems, but are said a little differently in the twenty-four-hour system, so that 10:00 am becomes "1000" (pronounced "ten-hundred") and 7:00 am becomes 0700 (pronounced "oh-seven-hundred" or "zero-seven-hundred"). Likewise, 3:38 am becomes 0338 ("oh-three-thirty-eight" or "zero-three-thirty-eight").

Midnight in the twenty-four-hour system is a little odd in that it can be expressed as 2400 or 0000, but one minute after midnight is always 0001, not 2401.

One final area of confusion. You may hear someone say "1300 hours" ("thirteen-hundred hours"). Technically this is not correct in the sea services (Navy, Marine Corps, and Coast Guard). The Army and Air Force append the word "hours," but Navy, Marine, and Coast Guard people should not. To be honest, this is one of those traditions that may be dying, so you may in fact hear Navy people saying "1722 hours" ("seventeen-twenty-two hours") instead of simply "1722."

Studying the table below should help you become more familiar with the twenty-four-hour system.

You may encounter the term "zero-dark-thirty" when talking with military people. This is just unofficial code for "very early."

Twenty-Four Hour Time System

Civilian time	Navy time	Spoken as
Midnight	0000 or 2400	"zero-zero-zero-zero" or "twenty-four hundred"
1 AM	0100	"Zero-one hundred" or "Oh-one-hundred"
2 AM	0200	"Zero-two-hundred" or" Oh-two-hundred"
3 AM	0300	
3:30 AM	0330	"Zero-three-thirty" or "Oh-three-thirty"
4:00 AM	0400	
5:00 AM	0500	
6:00 AM	0600	
6:15 AM	0615	"Zero-six-fifteen" or "Oh-six-fifteen"
6:16 AM	0616	"Zero-six-sixteen" or "Oh-six-sixteen"
7:00 AM	0700	
8:00 AM	0800	
9:00 AM	0900	
10:00 AM	1000	"Ten-hundred"
11:00 AM	1100	"Eleven-hundred"
11:47 AM	1147	"Eleven-forty-seven"
12 Noon	1200	
1:00 PM	1300	"Thirteen-hundred"
2:00 PM	1400	
3:00 PM	1500	
3:59 PM	1559	"Fifteen-fifty-nine"
4:00 PM	1600	
5:00 PM	1700	
6:00 PM	1800	
7:00 PM	1900	
8:00 PM	2000	"Twenty-hundred"
8:01 PM	2001	"Twenty-oh-one"
9:00 PM	2100	
10:00 PM	2200	
11:00 PM	2300	"Twenty-three-hundred"
12 Midnight	2400 or 0000	"Twenty-four-hundred" or "Zero-zero-zero-zero"
12:01 AM	0001	"Zero-zero-zero-one"

Dates

In the Navy, you will find that even dates are stated differently. Whereas the civilian convention is month-day-year (as in, "December 7, 1941" or "September 11, 2001"), in the Navy the format is day-month-year (as in, "7 December 1941" or "11 September 2001"). Further, you will often see an abbreviated version, such as "07AUG41" (note the use of the zero when the day of the month is less than ten) or "11SEP01."

On occasion you will see exceptions to this, as when a computer-friendly system is being used ("30 August 2007" becomes "20070830" for example) or other special circumstances, but the most frequently used format in the Navy is day-month-year.

"Minefields"

Just as a minefield should be avoided in war, so should you avoid wandering into certain areas of debate within the Navy. These are "battles" that are often waged among Navy people themselves, and unless you enjoy the gnashing of teeth and other such demonstrations of emotional behavior, you should probably leave them to their own private wars.

I am referring to the fact that there is a tendency among some Navy people (I must confess that I am one of them) to want to preserve tradition to the point of creating silly disputes that serve no useful purpose other than to exercise the blood pressure system among the participants.

I am referring to such raging debates as *whether the definite article belongs in front of a ship's name!* Traditionalists will insist (to the point of getting downright angry) that when one refers to a ship by name, it is improper to append the word "the" before the ship's name. In other words, one would "correctly" say, "*Constitution* is homeported in Boston," or, "*Independence* arrived off the coast before dawn," *not*, "The *Constitution* is homeported in Boston," or, "The *Independence* arrived off the coast before dawn." You can no doubt see the importance of this distinction; one can certainly see that the fate of the nation could well hang in the balance were one to overlook this convention! The truth is this tradition is all but dead—you will see it ignored more often than adhered to—but do not try to tell a traditionalist that it does not matter!

The use of feminine pronouns when referring to ships is a similar convention that is steeped in tradition and therefore not relinquished easily, but it, too, may well be dying a slow death.

Another area of some contention is the use of "in" versus "on" when discussing ships. When someone serves as part of a ship's official crew, the traditionally correct thing to say is, "She serves in USS *Arleigh Burke*," as opposed to, "She serves on USS *Arleigh Burke*." This practice dictates that one goes *on* a ship as a visitor but serves *in* a ship as part of the crew. This tradition is (in my estimation) more often ignored than practiced today, but it can have an effect on the impression you make, depending upon your audience.

Such things—and there are others—obviously do not affect the battle efficiency of our forces, but they can produce a surprising amount of passion, so it is best to avoid engagement. And do *not* poke fun as I have here—at least not without being very careful who your audience is. I am allowed because I am poking fun at myself. I am one of those fossils who has been known to get passionate about such things, despite my better judgment.

All said and done, it does not take a rocket scientist to figure out that if such things are important to your boss, they should be important to you. Or at least *appear* to be.

Safe Waters

There are some conventions that do not qualify as "minefields" that are worth your time to understand and use.

Ship Names

You may also notice that ship names are often *italicized* but the "USS" part is not. As already mentioned in the "Ships" chapter, this is a pretty standard convention, more often adhered to than not. But you should also be aware that in government circles you will also sometimes see a ship's name (and the USS) written in all caps, as in USS CONSTITUTION.

Ranks

As discussed in the "Military Titles" chapter, ranks are another area of confusion. In civilian publications, you will often see ranks abbreviated one way ("Lt. Cdr." for example), but in the Navy they are written a different way ("LCDR"). This stems from a variety of reasons: ignorance, the desire to limit the use of uppercase letters for aesthetic purposes, and so on. The only "rule" to follow is *know your audience.*

Uppercase versus Lowercase

You will see capitalization used more liberally in the military than in the civilian world. For example, a newspaper may refer to the "secretary of the navy" but a Navy directive will usually write "Secretary of the Navy."

Though there may be little or no logic to it (and little or no parallel in the civilian world—except perhaps the capitalization of "God" and associated pronouns, such as "He" and "Him"), there is an understood convention in the military that capitalization confers importance. In military culture, one instinctively capitalizes (1) whenever someone or something seems or is important, and (2) whenever in doubt. This latter tendency is the opposite of the conventions usually used in the civilian world, where capitalization is more often avoided. Consequently, from across a room you can usually identify a document as military because of the proliferation of uppercase letters on nearly every page.

The proliferation of acronyms does not help. Though our society in general is more and more prone to the use of acronyms, the military still holds the lead—by a wide margin—in the use of these efficient, but sometimes cryptic or exclusionary, "words."

And because some military communications systems archaically use all caps, you will sometimes find yourself reading a document that is written entirely in capital letters.

TWO VERY IMPORTANT TRADITIONS

Although we covered some of the Navy's many traditions in the "Navy Customs and Traditions" chapter, I saved two of the most important to close out this book. You are likely to encounter them no matter where you work within the Navy. They function as a reminder to those of us who serve this Navy that we are part of something important, not for its own sake but because it is one of the components that ensure the continued existence of this greatest nation in the history of the world. They can make us all walk just a bit taller and like what we see in the mirror because we are a part of the United States Navy and, when you reflect on it for just a moment, that really says quite a lot.

Friday the Thirteenth

On Friday, 13 October 1775, meeting in Philadelphia, the delegates to the Second Continental Congress voted to fit out two sailing vessels with ten carriage guns and eighty-man crews and sent them out on a cruise of three months to intercept transports carrying munitions and stores to the British army in America. This was an audacious move considering that it would be nearly nine months before that same Congress would produce the Declaration of Independence proclaiming the new United States of America. One might argue that it was also a move bordering on the irrational, considering the size and incredible power of the Royal Navy. One might also feel that Friday the thirteenth was not the ideal date to make such a move! But undaunted by such things, men like John Adams, who understood the importance of sea power, got the resolution passed, little realizing that they had signed the "birth certificate" of what would eventually become the most powerful navy in the history of the world.

In 1972, the Chief of Naval Operations issued a long-overdue decree, officially authorizing recognition of 13 October as the appropriate date for celebrating the Navy's birthday. Since that time, each CNO has encouraged a Navy-wide celebration of this occasion, "To enhance a greater appreciation of our Navy heritage, and to provide a positive influence toward pride and professionalism in the naval service." On that date, whether a Friday the thirteenth or some other day of the week, Sailors the world over gather for formal parties ashore, or have a piece of cake specially prepared by the cooks in the ship's galley, or simply pause for a moment before assuming their next watch to reflect on what it means to be part of an organization that is hundreds of years old, that has changed in countless ways, and yet carries with it the same bold spirit that was present at its birth.

Remembrance

Herman Wouk is one of the greatest writers of naval fiction. Once a Sailor himself, his Pulitzer Prize—winning *The Caine Mutiny* has been read by countless Sailors and has been often used as a textbook at the Naval Academy for its insights into a portion of the Navy's heritage and for its lessons on leadership. And Wouk's two books chronicling the history of World War II, *The Winds of War* and *War and Remembrance*, stand as monuments to that great struggle, not the least of which is his portrayal of the U.S. Navy in the victory at sea in the Pacific.

Although his works are technically fiction, they are filled with historical facts and, more importantly, they provide a deeper understanding of the human experience of war than any straight history can ever do. It is noteworthy then, that this great writer, only once in all the pages he wrote about the greatest conflict in history, stepped out of his role as anonymous narrator to make an overt comment about what he was writing.

He had been describing the sacrificial attack of the torpedo bombers at the Battle of Midway, telling how these aircraft, manned by an assortment of young men representing most of the states that are united to make up the USA, charged headlong into battle and certain death, and in so doing turned the tide of battle and altered the course of the world's history. He had been relating how, in just a very few minutes, thirty-three pilots and thirty-five radiomen-gunners were killed as they distracted enemy fighter aircraft long enough for their fellow Sailors in the dive bombers to strike a fatal blow into the very heart of the Imperial Japanese Navy, the same aircraft carriers that had struck at Pearl Harbor on the first day of the war. Describing this incredible moment as "the soul of the United States of America in action," Wouk wrote, "The memory of these three American torpedo plane squadrons should not die." He then halted the telling of his story, pausing for several pages to list the names and the birthplaces of each of the sixty-eight men who paid the ultimate price that day.

This unusual tribute to real people in the midst of a work of fiction tells us that there is something truly extraordinary about the Battle of Midway. The Naval Historical Center, in describing the battle on its Web site, calls it, "The decisive battle of the war in the Pacific," and few historians dispute that. It was a battle that was won by the courage and sacrifice of those pilots and gunners who took to the air knowing that the odds were stacked against them. It was also won through some exceptional intelligence work by a handful of military and civilian cryptanalysts working long hours for weeks on end to decipher the enemy's intentions. And (as mentioned in the introduction to this book) by hundreds of civilian shipyard workers who performed maintenance miracles to ensure that there were three American aircraft carriers available to fight instead of only two. And by several admirals who made key decisions at the right times. And by those anonymous others who kept the Navy running ashore, who built those great ships in the first place, who maintained the vital logistics trail, who muscled the bombs onto aircraft, found the winds for the launch, kept the steam flowing and the electrons streaming, prepared the meals, kept the records, cleaned, lubricated, repaired, and performed countless other duties to ensure that this outnumbered and outclassed fleet was ready to do battle, to do what U.S. Sailors are meant to do.

Recognizing the particular significance of this battle and, more importantly, what it represents—the efforts of men and women all over the world merged into a great culminating moment that changed the course of the war and the history of the world—the CNO sent out a message to the entire Navy on the fifty-eighth anniversary of the battle, proclaiming:

> The two most significant dates in our Navy's history are *13 October 1775, the birth of our Navy*, and *4 June 1942, the Battle of Midway*. These two prominent days will henceforth be celebrated annually as the centerpieces of our heritage. Twice a year, we will pause as a Navy, to reflect upon our proud heritage and to build in all hands a renewed awareness of our tradition and history. Through such reflection, we will help define the significance of our service today in defense of our country's freedom. We are the caretakers of a torch passed on by nearly two hundred and twenty five years of naval heroes. Each of us will be enriched by honoring their contributions. I believe it is appropriate that we take time to pause and reflect formally upon our proud naval heritage.

And, ever since, the Navy has heeded the CNO's words. In the spring of every year, Navy people the world over gather together to remember a moment in our Navy's history when the extraordinary was accomplished by the ordinary, to reflect on what that represents to each man and woman who plays a role in making this Navy the greatest the world has ever seen. It is a moment when one excuses the many idiosyncrasies that make this organization difficult to understand at times, when it becomes clear that unusual things are not often accomplished in usual ways, when one allows a little self-indulgent smile and feels *special*.

QUICKREFS

- Numbers are often pronounced individually: "one, zero, seven" rather than "one hundred seven," and "nine" is pronounced "niner" to distinguish it from "five" on a radio or telephone circuit.
- The phonetic alphabet is often used in the Navy:

Alfa
Bravo
Charlie
Delta
Echo
Foxtrot
Golf
Hotel
India
Juliett
Kilo
Lima
Mike
November
Oscar
Papa
Quebec
Romeo
Sierra
Tango
Uniform
Victor
Whiskey
Xray
Yankee
Zulu

- "Roger" (I heard you) and "Wilco" (I heard and understood you and will do what you say) should never be used together. One or the other.

- "Over" (I am finished talking and now it's your turn to talk) and "Out" (I am finished talking and this conversation is over) should never be used together.
- The term "Bravo Zulu" is a traditional term derived from the phonetic alphabet and from Navy codebooks that means "well done."
- The Navy uses the more efficient *twenty-four-hour time system* instead of the twelve-hour (am and pm) system used in the civilian world. Instead of repeating the twelve-hour cycle, the twenty-four-hour system merely continues, so that 1:00 pm is instead 1300, 2:00 pm is 1400, and so on. No colons are used and zeros are added where necessary so that time is always expressed in four digits (0800, 0407, 0930, etc.). Times are pronounced as "oh-eight-hundred" or "zero-eight-hundred" (0800), "ten-thirty-seven" (1037), "fifteen-forty-one" (1541), and so on. The word "hours" (as in "thirteen-hundred hours") is used in the Army and Air Force but not used in the sea services.
- *Dates* in the Navy are usually expressed in day-month-year format, as in "11 October 2007" and are often abbreviated, as in "11OCT07."
- Be aware that there are continuing arguments in the Navy that some people take very seriously, such as whether a ship's name should be preceded by the definite article (as in "the USS *Enterprise*" versus just "USS *Enterprise*").
- *Ship names* usually appear in italics or all caps, as in "USS *Arleigh Burke*" or "USS ARLEIGH BURKE."
- *Ranks* are abbreviated in different ways depending upon the usage. There is a military way (LT) and a civilian way (Lt.).
- *Capitalization* is used more frequently in the military world than in the civilian. It is often used to convey importance, as in Secretary of the Navy.
- The *Navy birthday* is on 13 October (1775 is the original year) and is celebrated throughout the Navy in various ways.
- The *Battle of Midway* is commemorated every year in the Navy on or about its anniversary (the battle was fought over several days but is usually considered to be 4 June 1942).

Selected Sources and Resources

BOOKS

The suggested list below may seem nepotistic because all of the books are from Naval Institute Press, but the truth is that this organization that I am privileged to work for is *the* publisher for professional books related to the sea services. It is indeed one of the primary reasons why the Naval Institute has existed since 1873.

The Naval Institute Almanac of the U.S. Navy by Anthony Cowden packs a lot of useful information into a small package. A look at the current Navy is supplemented by a lot of timeless information about the Navy. Included are a number of useful reference lists, such as the names of all the Secretaries of the Navy that have held that post and a list of major Navy bases. A particularly useful feature of this book is a list of naval-oriented Web sites.

Much of what is in this book is replicated in *The Bluejacket's Manual* (twenty-third edition) by Thomas J. Cutler, so you need not obtain a copy unless you are seeking more detail and a number of topics not covered here. This is the book that every new recruit receives upon entering the U.S. Navy and is both an introduction to the Navy and a reference book that most Sailors have kept handy for more than a century. You might have noted that I am the author of the book, but it should be clearly understood that I have not been writing it since it first appeared in 1902!

In "The Paper Navy" (chapter 4), I pointed out that Robert Shenk's *The Naval Institute Guide to Naval Writing* is an excellent aid to writing all kinds of naval documents, from memos to formal directives. As of this writing, Bob is working on a third edition that will be even more useful (and up-to-date) than the previous two, which have been very well received.

Two reference books that I must mention because they are considered very useful volumes to keep handy at any Navy desk (and because I prefer to remain married) are the *Dictionary of Naval Terms* and the *Dictionary of Naval Abbreviations*, both of which are by Deborah W. Cutler and Thomas J. Cutler.

For more information than would fit into chapter 5 ("Navy Customs and Traditions"), see *Naval Ceremonies, Customs, and Traditions* (sixth edition) by Royal W. Connell and William P. Mack. This book has long been the "bible" for those who would have a better understanding of this wonderful (but odd) organization we are a part of.

PERIODICALS

Nepotism continues as I recommend U.S. Naval Institute *Proceedings* magazine as a means of staying up on the *dialogue* regarding sea service matters. This periodical is different from many others in that it has no standing point of view. It is not limited to official policy and serves as an open forum where Sailors, civilians, contractors, citizens, and so on can voice their views, introduce their ideas, share their thinking, constructively criticize policies, and offer alternatives. The only guiding criteria as to what goes into this magazine is that it should ultimately serve the greater good of the Navy. The annual "Naval Review" issue (always published in May) is a larger issue packed with additional analysis and reference information.

Navy Times is the unofficial but well-informed weekly newspaper that provides the latest news on what is happening or what is being contemplated in the Navy. Like *Proceedings* it is an independent voice that is not limited by official policy. It offers a wider variety of information that includes promotion lists, entertainment features, the latest pay scales, and so on.

Always geared toward the services' best interests, the Navy League's *Seapower* magazine is an excellent source for keeping up on what is happening in technology, policy, leadership, and other issues within the sea services. The annual (January issue) *Seapower* Almanac is worth the (very reasonable) price of membership; filled with excellent analysis and handy reference information, you will do well to keep it handy throughout the year.

The Navy's official *All Hands* magazine serves as a good familiarization— a kind of visual tour—with what is happening in the Navy. Features focus on the activities of Sailors around the world doing ordinary and sometimes extraordinary things. You will not find much analysis but some of the articles are very enlightening (such as the story of how Sailors in USS *Cole* dealt with the aftermath of the infamous terrorist attack in Yemen a year before 9/11). The January issue—called "Owner's and Operator's Manual"—is another excellent source of reference information that should be kept handy year-round. In this special issue you will find quick reference features on ships, air-craft, weapons, ranks, ribbons, and a number of other items that make this a worthwhile addition to a briefcase or a desktop. It is also available online (see below).

WEB SITES

As mentioned above, the *All Hands* "Owner's and Operator's Manual" is available online at http://www.news.navy.mil/allhands.asp. Once at this site, you can select the issue you want (go to January for the latest "Owner's and Operator's Manual").

To get the latest official information on what is happening in the Navy, go to http://www.navy.mil. There you will find links to such things as "Status of the Navy," which provides up-to-date information on how many personnel are currently in the Navy, where ships currently are, the latest Navy news, and so on. You will also find links to many other useful Web sites.

At http://doni.daps.dla.mil you will get access to virtually all of the direc-tives issued by the Secretary of the Navy and the Chief of Naval Operations. You will also find links to various manuals and to the "Standard Navy Distri-bution List," which provides mailing addresses to all Navy commands. There is also a button called "Subscribe to New Directives" that allows you to be notified by e-mail each time a new directive is issued or an old one is modified or deleted. This is an excellent way to keep up on policy changes and the like.

At http://www.npc.navy.mil/ReferenceLibrary/Messages you will gain access to the latest Navy messages issued by the Secretary of the Navy (ALNAVs) and those issued by the Chief of Naval Operations (NAVADMINs).

The latest official Navy news can be obtained at http://www.news.navy.mil/index.asp and the unofficial news can be obtained at http://www.navytimes.com (although to get the full benefit of this Web site you need to subscribe).

Another source of unofficial information (less timely than what can be found at the news sites above but more analytical and in-depth) is the Naval Institute's Web site at http://www.navalinstitute.org. Like the *Navy Times* Web site, to get the full benefit you have to subscribe.

To find the official status of any or all of the Navy's ships, go to the Naval Vessel Register at http://www.nvr.navy.mil. There you will find up-to-date information on fleet size, ship names, hull classifications, and so on.

Another excellent source of information about ships is the Dictionary of American Naval Fighting Ships at http://www.history.navy.mil/danfs/index.html. Here you will find historical information on naval warships.

For a wealth of other historical information, go to the Naval Historical Center's Web site at http://www.history.navy.mil.

The Defense Technical Information Center Web site at http://www.dtic.mil is a good site to search for more detailed information on a particular topic that has appeared in reports, official journals, and so on.

Acronyms and Abbreviations

Commonly used ship-type abbreviations (such as "CV" for aircraft carrier) are included here, but those less often encountered can be found in chapter 6 ("Ships").

1LT	First Lieutenant (Army)
1MC	general announcing system
1SG	First Sergeant (Army)
1st Lt	First Lieutenant (Air Force)
1stLt	First Lieutenant (Marine Corps)
1stSgt	First Sergeant (Marine Corps)
2d Lt	Second Lieutenant (Air Force)
2LT	Second Lieutenant (Army)
2ndLt	Second Lieutenant (Marine Corps)
A1C	Airman First Class (Air Force)
AA	Airman Apprentice (Navy and Coast Guard)
AARGM	advanced anti-radiation guided missile
AAW	anti-air warfare
AB	Airman Basic (Air Force)
ADM	Admiral (Navy and Coast Guard)
AE	ammunition ship
AH	hospital ship
AIRLANT	Naval Air Force, U.S. Atlantic Fleet
AIRPAC	Naval Air Force, U.S. Pacific Fleet
ALNAV	all of the Navy (and the Marine Corps)
ALWT	advanced lightweight torpedo

AMC	Air Mobility Command (Air Force)
Amn	Airman (Air Force)
AMRAAM	advanced medium-range air-to-air missile
AN	Airman (Navy and Coast Guard); Army Navy (JETDS prefix)
AO	oiler
AOR	Area of Responsibility; replenishment oiler
AR	Airman Recruit (Navy and Coast Guard)
AS	submarine tender
ASROC	antisubmarine rocket
ASUW	antisurface warfare
ASW	antisubmarine warfare
AW	air warfare
BB	battleship
BG	Brigadier General (Army)
BGen	Brigadier General (Marine Corps)
Brig Gen	Brigadier General (Air Force)
BUMED	Bureau of Medicine and Surgery
BUPERS	Bureau of Naval Personnel
BZ	Bravo Zulu ("well done")
CA	Constructionman Apprentice
CAP	combat air patrol
Capt	Captain (Air Force and Marine Corps)
CAPT	Captain (Navy and Coast Guard)
CB	SeaBee (Construction Battalion)
CBR	chemical, biological, or radiation (attack)
CC	Command Chief Petty Officer (Coast Guard)
CDR	Commander (Navy and Coast Guard)
CENTCOM	Central Command
CFFC	Commander Fleet Forces Command
CG	guided-missile cruiser
CHC	Chaplain Corps
CIC	combat information center
CIWS	close-in weapons system
CJCS	Chairman of the Joint Chiefs of Staff
CMC	Command Master Chief Petty Officer Command Master Chief Petty Officer (Coast Guard)

CMDCM	Command Master Chief Petty Officer
CMSAF	Chief Master Sergeant of the Air Force
CMSgt	Chief Master Sergeant (Air Force)
CN	Constructionman
CNO	Chief of Naval Operations
CNOCM	CNO-Directed Command Master Chief Petty Officer
Col	Colonel (Air Force and Marine Corps)
COL	Colonel (Army)
COMSEC	secure communications
COMLANTFLT	Commander of the Atlantic Fleet
COMNAVCRUITCOM	Commander Navy Recruiting Command
COMNAVSURFLANT	Commander Naval Surface Forces Atlantic
COMNAVSURFPAC	Commander Naval Surface Forces Pacific
COMSECONDFLT	Commander of the Second Fleet
COMSIXTHFLT	Commander of the Sixth Fleet
COMUSEUCOM	Commander U.S. European Command
COMUSNAVEUR	Commander of U.S. Naval Forces in Europe
Cpl	Corporal (Marine Corps)
CPL	Corporal (Army)
CPO	Chief Petty Officer (Navy and Coast Guard)
CPT	Captain (Army)
CR	Constructionman Recruit
CSC	Command Senior Chief Petty Officer (Coast Guard)
CSM	Command Sergeant Major (Army)
CTE	Commander Task Element
CTF	Commander Task Force
CTG	Commander Task Group
CTU	Commander Task Unit
CV	aircraft carrier
CVW	carrier air wing
CW2	Chief Warrant Officer (W-2) (Army)
CW3	Chief Warrant Officer (W-3) (Army)
CW4	Chief Warrant Officer (W-4) (Army)
CW5	Chief Warrant Officer (W-5) (Army)
CWO2	Chief Warrant Officer (W-2) (Navy, Coast Guard, and Marine Corps)

CWO3	Chief Warrant Officer (W-3) (Navy, Coast Guard, and Marine Corps)
CWO4	Chief Warrant Officer (W-4) (Navy, Coast Guard, and Marine Corps)
CWO5	Chief Warrant Officer (W-5) (Navy and Marine Corps)
DCC	damage control central
DCNO	Deputy Chief of Naval Operations
DDG	guided-missile destroyer
DESRON	destroyer squadron
DOD	Department of Defense
DODPM	Department of Defense Military Pay and Allowance Entitlements Manual
DON	Department of the Navy
DTG	Date-Time Group
ECCM	electronic counter-countermeasures
ECM	electronic countermeasures
EMCON	emission control
Encl	enclosure (component of a naval letter)
ENS	Ensign (Navy and Coast Guard)
EPM	electronic protective measures
ERGM	Extended Range Guided Munition
ESM	electronic warfare support measures
EUCOM	European Command
EW	electronic warfare
FA	Fireman Apprentice (Navy and Coast Guard)
FADM	Fleet Admiral
FFG	guided-missile frigate
FN	Fireman (Navy and Coast Guard)
FLTCM	Fleet Master Chief Petty Officer
FM	from
FORCM	Force Master Chief Petty Officer
FOUO	for official use only
FPO	Fleet Post Office
FR	Fireman Recruit (Navy and Coast Guard)
FUSPAD	*Forward—Up—Starboard, Port—Aft—Down*
GBU	Guided Bomb Unit

Gen	General (Air Force and Marine Corps)
GEN	General (Army)
GP	general purpose (bomb)
GPO	Government Printing Office
GQ	General Quarters
GS	General Schedule
GySgt	Gunnery Sergeant (Marine Corps)
H	helicopter
HA	Hospitalman Apprentice
HARM	high-speed anti-radiation missile
HC	helicopter combat support (squadron)
HCS	helicopter combat support, special (squadron)
HM	helicopter mine countermeasures (squadron)
HMS	Her (His) Majesty's Ship (Royal Navy)
HN	Hospitalman
HR	Hospitalman Recruit
HS	helicopter antisubmarine (squadron)
HSL	light helicopter antisubmarine (squadron)
HT	helicopter training (squadron)
ICBM	intercontinental ballistic missile
IFF	identification friend or foe
INFO	information addressee(s)
INST	instruction (as in SECNAVINST for Secretary of the Navy Instruction)
IRBM	intermediate-range ballistic missile
ISR	intelligence, surveillance, and reconnaissance
JAG	Judge Advocate General
JAGMAN	Manual of the Judge Advocate General
JCSJoint	Chiefs of Staff
JDAM	joint direct attack munition
JETDS	Joint Electronics Type Designation System
JFCOM	Joint Forces Command
JFTR	Joint Federal Travel Regulations
JOOD	Junior Officer of the Deck; Junior Officer of the Day
LAMPS	Light Airborne Multipurpose System (ship-board helicopter)

LANTCOM	Atlantic Command (archaic)
LCAC	air-cushioned landing craft
LCDR	Lieutenant Commander (Navy and Coast Guard)
LCpl	Lance Corporal (Marine Corps)
LDGP	low-drag, general purpose (bomb)
LGB	laser-guided bomb
LGP	laser-guided bomb
LHA	amphibious assault ship (general purpose)
LHD	amphibious assault ship (multipurpose)
LPD	amphibious transport dock (ship)
LSD	dock landing ship
LT	Lieutenant (Navy and Coast Guard)
LTC	Lieutenant Colonel (Army)
Lt Col	Lieutenant Colonel (Air Force)
LtCol	Lieutenant Colonel (Marine Corps)
LTG	Lieutenant General (Army)
Lt Gen	Lieutenant General (Air Force)
LtGen	Lieutenant General (Marine Corps)
LTJG	Lieutenant (Junior Grade) (Navy and Coast Guard)
MAA	Master-at-Arms
MAC	Military Airlift Command (Air Force) (archaic)
Mach	speed of sound
Maj	Major (Air Force and Marine Corps)
MAJ	Major (Army)
Maj Gen	Major General (Air Force)
MajGen	Major General (Marine Corps)
MCM	Manual for Courts-Martial
MCPO	Master Chief Petty Officer (Navy and Coast Guard)
MCPOCG	Master Chief Petty Officer of the Coast Guard
MCPON	Master Chief Petty Officer of the Navy
MDS	Mission Design Series (aircraft)
MEDEVAC	medical evacuation
MEU	Marine Expeditionary Unit

MG	Major General (Army)
MGySgt	Master Gunnery Sergeant (Marine Corps)
MILPERSMAN	Naval Military Personnel Manual
MIRV	multiple independently targetable reentry vehicles
MIW	mine warfare
MOD	modification
MOS	Military Occupational Specialty (Army, Air Force, Marine Corps)
MSC	Military Sealift Command
MSG	Master Sergeant (Army)
MSgt	Master Sergeant (Air Force and Marine Corps)
MUC	Meritorious Unit Commendation
M/BVR	medium/beyond visual range
NATO	North Atlantic Treaty Organization
NAVADMIN	administrative message from the CNO
NAVAIR	Naval Air Systems Command
NAVCENT	Naval Forces Central Command
NAVCOMP	Comptroller of the Navy
NAVEDTRA	Naval Education and Training
NAVEUR	Naval Forces Europe
NAVFAC	Naval Facilities Engineering Command
NAVMAT	Naval Material Command
NAVMILPERSCOM	Naval Military Personnel Command
NAVREGS	U.S. Navy Regulations
NAVSEA	Naval Sea Systems Command
NAVSO	Navy Staff Office (Executive Offices of the Secretary of the Navy)
NAVSUP	Naval Supply Systems Command
NCIS	Naval Criminal Investigative Service
NEDS	Navy Electronic Directives System
NETC	Naval Education and Training Command
NFAF	Naval Fleet Auxiliary Force
NFO	Naval Flight Officer
NJP	nonjudicial punishment
NORAD	North American Aerospace Command

NORTHCOM	Northern Command
NOTE	notice (as in OPNAVNOTE 5400)
NUC	Navy Unit Commendation
NWTD	non-watertight door
OOD	Officer of the Deck; Officer of the Day
OPNAV	Office of the Chief of Naval Operations
PACFLT	Pacific Fleet
PACOM	Pacific Command
PAYPERSMAN	Navy Pay and Personnel Procedures Manual
PFC	Private First Class (Army and Marine Corps)
PO1	Petty Officer First Class (Navy and Coast Guard)
PO2	Petty Officer Second Class (Navy and Coast Guard)
PO3	Petty Officer Third Class (Navy and Coast Guard)
POA&M	plan of action and milestones
POD	Plan of the Day
POS	protection of shipping
PREP	preparatory (pennant)
PSG	Platoon Sergeant (Army)
PV2	Private Second Class (Army)
PUC	Presidential Unit Citation
Pvt	Private (Marine Corps)
PVT	Private (Army)
RADHAZ	radiation hazard
RADIAC	radioactivity detection, identification, and computation
RADM	Rear Admiral (Upper Half) (Navy and Coast Guard)
RAM	Rolling Airframe Missile
RDML	Rear Admiral (Lower Half) (Navy and Coast Guard)
Ref	reference (component of a naval letter)
SA	Seaman Apprentice (Navy and Coast Guard)
SAC	Strategic Air Command (Air Force) (archaic)
SACEUR	Supreme Allied Command Europe
SAP	semi-armor-piercing (bomb)

SC	Supply Corps
SCPO	Senior Chief Petty Officer (Navy and Coast Guard)
SDDC	Surface Deployment and Distribution Command (Army)
SeaBee	Construction Battalion (CB)
SEAL	Sea Air Land (Naval Special Forces)
SECDEF	Secretary of Defense
SECNAV	Secretary of the Navy
SFC	Sergeant First Class (Army)
SGM	Sergeant Major (Army)
Sgt	Sergeant (Marine Corps)
SGT	Sergeant (Army)
SgtMaj	Sergeant Major; Sergeant Major of the Marine Corps (Marine Corps)
SLAM	Standoff Land Attack Missile
SMA	Sergeant Major of the Army
SMSgt	Senior Master Sergeant (Air Force)
SN	Seaman (Navy and Coast Guard)
SNDL	Standard Navy Distribution List
SOCOM	Special Operations Command
SORM	Standard Organization and Regulations of the U.S. Navy
SOUTHCOM	Southern Command
SP	Shore Patrol
SP4	Specialist (E-4) (Army)
SPACECOM	Space Command (archaic)
SPAWAR	Space and Naval Warfare Systems Command
SR	Seaman Recruit (Navy and Coast Guard)
SrA	Senior Airman (Air Force)
SS	submarine warfare
SSBN	ballistic-missile submarine (nuclear powered)
SSG	Staff Sergeant (Army)
SSGN	guided-missile submarine (nuclear powered)
SSN	attack submarine (nuclear powered)
SSgt	Staff Sergeant (Air Force and Marine Corps)
SSIC	Standard Subject Identification Code
STRATCOM	Strategic Command

Subj	subject (component of a naval letter)
SUBLANT	Naval Submarine Force, U.S. Atlantic Fleet
SUBPAC	Naval Submarine Force, U.S. Pacific Fleet
SURFLANT	Naval Surface Force, U.S. Atlantic Fleet
SURFPAC	Naval Surface Force, U.S. Pacific Fleet
SW	surface warfare
TASM	Tomahawk anti-ship missile
TE	Task Element
TF	Task Force
TG	Task Group
TLAM	Tomahawk land-attack missile
TLAM(N)Tomahawk	land-attack missile (nuclear)
TRANSCOM	Transportation Command
TRANSMAN	Enlisted Transfer Manual
TSgt	Technical Sergeant (Air Force)
TU	Task Unit
UCMJ	Uniform Code of Military Justice
USA	U.S. Army
USACOM	U.S. Atlantic Command (archaic)
USAF	U.S. Air Force
USARCENT	U.S. Army Central Command
UAV	unmanned aerial vehicle
UNCLAS	unclassified
UNREP	underway replenishment
USCENTAF	U.S. Central Command Air Forces
USCENTCOM	U.S. Central Command
USCG	U.S. Coast Guard
USCGC	United States Coast Guard Cutter
USEUCOM	U.S. European Command
USJFCOM	U.S. Joint Forces Command
USMARCENT	U.S. Marine Forces Central Command
USMC	U.S. Marine Corps
USN	U.S. Navy
USNAVCENT	U.S. Naval Forces Central Command
USNAVEUR	U.S. Naval Forces Europe
USNAVSO	U.S. Naval Forces Southern Command
USNORTHCOM	U.S. Northern Command
USNS	United States Naval Ship (MSC)

USPACFLT	U.S. Pacific Fleet
USPACOM	U.S. Pacific Command
USS	United States Ship
USSOCOM	U.S. Special Operations Command
USSOUTHCOM	U.S. Southern Command
USSTRATCOM	U.S. Strategic Command
USTRANSCOM	U.S. Transportation Command
USW	undersea warfare
UTC	Coordinated Universal Time
VA	Department of Veterans Affairs (Veterans Administration); attack (aviation squadron)
VADM	Vice Admiral (Navy and Coast Guard)
VAQ	tactical electronic warfare (aviation squadron)
VAW	carrier airborne early warning (aviation squadron)
VC	fleet composite (aviation squadron)
VCNO	Vice Chief of Naval Operations
VERTREP	vertical replenishment
VF	fighter (aviation squadron)
VFA	strike fighter (aviation squadron)
VFC	fighter composite (aviation squadron)
VLS	vertical launch system (missiles)
VP	patrol (aviation squadron)
VQ	reconnaissance/strategic communications (aviation squadron)
VR	fleet logistics support (aviation squadron)
VRC	carrier logistics support (aviation squadron)
VS	sea control (aviation squadron)
V/STOL	vertical/short take off and landing
VT	training (aviation squadron)
VX	test and evaluation (aviation squadron)
WESTPAC	Western Pacific
WO	Warrant Officer (Marine Corps)
WO1	Warrant Officer (W-1) (Army)
WTD	watertight door
Z	Zulu; Coordinated Universal Time (as part of a DTG); flash (precedence on a message)

Index

Note: The prefix United States (U.S.) has been omitted from formal titles.

About the Author

Thomas J. Cutler is a retired lieutenant commander and former gunner's mate second class who served in patrol craft, cruisers, destroyers, and aircraft carriers. His varied assignments included an in-country Vietnam tour, small craft command, and nine years at the U.S. Naval Academy, where he served as executive assistant to the chairman of the Seamanship and Navigation Department and associate chairman of the History Department. While at the Academy, he was awarded the William P. Clements Award for Excellence in Education (military teacher of the year).

He is the founder and former director of the Walbrook Maritime Academy in Baltimore. Currently he is Fleet Professor of Strategy and Policy with the Naval War College, teaches Naval Warfare at the U.S. Naval Academy, and is the Director of Professional Publishing at the U.S. Naval Institute.

Winner of the Alfred Thayer Mahan Award for Naval Literature and the U.S. Maritime Literature Award, his published works include *A Sailor's History of the U.S. Navy* (Naval Institute Press/Naval Historical Center, 2005), which is currently issued to all Sailors entering the Navy, *The Battle of Leyte Gulf* (HarperCollins, 1994), and *Brown Water, Black Berets: Coastal & Riverine Warfare in Vietnam* (Naval Institute Press, 1988). His books have been published in various forms, including paperback and audio, and have appeared as main and alternate selections of the History Book Club, Military Book Club, and Book of the Month Club. He is the author of the twenty-second and twenty-third (Centennial) editions of *The Bluejacket's Manual,* and his other works include revisions of Jack Sweetman's *The Illustrated History of the U.S. Naval Academy* and *Dutton's Nautical Navigation.* He and his wife, Deborah W. Cutler, are the coeditors of the *Dictionary of Naval Terms* and the *Dictionary of Naval Abbreviations.*

He has served as a panelist, commentator, and keynote speaker on military and writing topics at many events and for various organizations, including the Naval Historical Center, Smithsonian Institution, Navy Memorial, U.S. Naval Academy, MacArthur Memorial Foundation, Johns Hopkins University, U.S. Naval Institute, Armed Forces Electronics Communications and Electronics Association, Naval War College, Civitan, and many veterans' organizations. His television appearances include the History Channel's *Biography* series, A&E's *Our Century*, Fox News Channel's *The O'Reilly Factor*, and CBS's *48 Hours*.

The Naval Institute Press is the book-publishing arm of the U.S. Naval Institute, a private, nonprofit, membership society for sea service professionals and others who share an interest in naval and maritime affairs. Established in 1873 at the U.S. Naval Academy in Annapolis, Maryland, where its offices remain today, the Naval Institute has members worldwide.

Members of the Naval Institute support the education programs of the society and receive the influential monthly magazine *Proceedings* or the colorful bimonthly magazine *Naval History* and discounts on fine nautical prints and on ship and aircraft photos. They also have access to the transcripts of the Institute's Oral History Program and get discounted admission to any of the Institute-sponsored seminars offered around the country.

The Naval Institute's book-publishing program, begun in 1898 with basic guides to naval practices, has broadened its scope to include books of more general interest. Now the Naval Institute Press publishes about seventy titles each year, ranging from how-to books on boating and navigation to battle histories, biographies, ship and aircraft guides, and novels. Institute members receive significant discounts on the Press's more than eight hundred books in print.

Full-time students are eligible for special half-price membership rates. Life memberships are also available.

For a free catalog describing Naval Institute Press books currently available, and for further information about joining the U.S. Naval Institute, please write to:

Member Services
U.S. Naval Institute
291 Wood Road
Annapolis, MD 21402–5034
Telephone: (800) 233–8764
Fax: (410) 571–1703
Web address: www.usni.org